Learn Data Mining Through Excel

A Step-by-Step Approach for Understanding Machine Learning Methods

Second Edition

Hong Zhou

Apress®

Learn Data Mining Through Excel: A Step-by-Step Approach for Understanding Machine Learning Methods

Hong Zhou
Department of Mathematics and Computer Science
University of Saint Joseph, West Hartford, CT, USA

ISBN-13 (pbk): 978-1-4842-9770-4   ISBN-13 (electronic): 978-1-4842-9771-1
https://doi.org/10.1007/978-1-4842-9771-1

Managing Director, Apress Media LLC: Welmoed Spahr
Acquisitions Editor: Mark Powers
Development Editor: Laura Berendson
Editorial Project Manager: Shaul Elson

Cover designed by eStudioCalamar

Cover image by Vecstock on Freepik (www.freepik.com)

Distributed to the book trade worldwide by Springer Science+Business Media New York, 1 New York Plaza, Suite 4600, New York, NY 10004-1562, USA. Phone 1-800-SPRINGER, fax (201) 348-4505, e-mail orders-ny@springer-sbm.com, or visit www.springeronline.com. Apress Media, LLC is a California LLC and the sole member (owner) is Springer Science + Business Media Finance Inc (SSBM Finance Inc). SSBM Finance Inc is a **Delaware** corporation.

For information on translations, please e-mail booktranslations@springernature.com; for reprint, paperback, or audio rights, please e-mail bookpermissions@springernature.com.

Apress titles may be purchased in bulk for academic, corporate, or promotional use. eBook versions and licenses are also available for most titles. For more information, reference our Print and eBook Bulk Sales web page at http://www.apress.com/bulk-sales.

Any source code or other supplementary material referenced by the author in this book is available to readers on GitHub (github.com/apress). For more detailed information, please visit https://www.apress.com/gp/services/source-code.

Paper in this product is recyclable.

For my family and friends.

Table of Contents

About the Author

Hong Zhou, PhD, is a professor of computer science and mathematics and has been teaching courses in computer science, data science, mathematics, and informatics at the University of Saint Joseph for nearly 20 years. His research interests include bioinformatics, data mining, software agents, and blockchain. Prior to his current position, he was a Java developer in Silicon Valley. Dr. Zhou believes that learners can develop a better foundation of data mining models when they visually experience them step by step, which is what Excel offers. He has employed Excel in teaching data mining and finds it an effective approach for both data mining learners and educators.

About the Technical Reviewer

Adam Gladstone has over 25 years' experience in software development, mostly in C++ and C#. He has worked mainly in the investment banking and finance sectors. For the last few years, he has been developing data science and machine learning skills, particularly in Python and R after completing a degree in maths and statistics. He loves programming in C++ and C# and his free time is spent developing software tools.

CHAPTER 1

Excel and Data Mining

Let's get right to the topic. Why do we need to learn Excel in our data mining endeavor? It is true that there are some outstanding data mining software tools such as RapidMiner that make the mining process easy and straightforward. In addition, programming languages Python and R have a large number of reliable packages dedicated to various data mining tasks. What is the purpose of studying data mining or machine learning through Excel?

Why Excel?

If you are already an experienced data mining professional, I would say that you are asking the right question and probably you should not read this book. However, if you are a beginner in data mining, or a visual learner, or want to understand the mathematical background behind some popular data mining techniques, or an educator, then this book is right for you, and probably is the first book you should read before you start your data mining journey.

Excel allows you to work with data in a transparent manner, meaning when an Excel file is opened, the data is visible immediately and every step of data processing is also visible. Intermediate results are contained in the Excel worksheet and can be examined while you are conducting your mining task. This allows you to obtain a deep and clear understanding of how the data are manipulated and how the results are obtained. Other software tools and programming languages hide critical aspects of the model construction process. For most data mining projects, the goal is to find the internal hidden patterns inside the data. Therefore, hiding the detailed process is beneficial to the users of the tools or packages. But it is not helpful for beginners, visual learners, or those who want to understand how the mining process works. Let me use k-nearest neighbors method (K-NN) to illustrate the learning differences between RapidMiner, R, and Excel. Before we do that, we need to understand several terminologies in data mining.

1

© Hong Zhou 2023
H. Zhou, *Learn Data Mining Through Excel*, https://doi.org/10.1007/978-1-4842-9771-1_1

There are two types of data mining techniques: supervised and unsupervised. Supervised methods require the use of a training dataset to "train" the software programs or algorithms (such programs or algorithms are often referred to as machines) first. Programs are trained to reach an optimal state called a model. This is why a training process is also called modeling. Data mining methods can also be categorized into parametric and nonparametric methods. For parametric methods, a model is just a set of parameters or rules obtained through the training process that are believed to allow the programs to work well with the training dataset. Nonparametric methods do not generate a set of parameters. Instead, they dynamically evaluate the incoming data based on the existing dataset. You may be confused by such definitions at this time. They will make sense soon.

What is a training dataset? In a training dataset, the target variable (also called label, target, dependent variable, outcome variable, response), the value of which is to be predicted, is given or known. The value of the target variable depends on the values of other variables which are usually called attributes, predictors, or independent variables. Based on the attribute values, a supervised data mining method computes (or so-called predicts) the value of the target variable. Some computed target values might not match the known target values in the training dataset. A good model indicates an optimal set of parameters or rules that can minimize the mismatches.

A model is usually constructed to work on future datasets with unknown target values in a supervised data mining method. Such future datasets are commonly called scoring datasets. In an unsupervised data mining method, however, there is no training dataset and the model is an algorithm that can directly be applied on the scoring datasets. K-nearest neighbors method is a supervised data mining technique.

Suppose we want to predict if a person is likely to accept a credit card offer based on the person's age, gender, income, and number of credit cards they already have. The target variable is the response to the credit card offer (assume it is either Yes or No), while age, gender, income, and number of existing credit cards are the attributes. In the training dataset, all variables including both the target and attributes are known. In such a scenario, a K-NN model is constructed through the use of the training dataset. Based on the constructed model, we can predict the responses to the credit card offer of people whose information is stored in the scoring dataset.

In RapidMiner, one of the best data mining tools, the prediction process is as follows: retrieve both the training data and scoring data from the repository ➤ set role for the training data ➤ apply the K-NN operator on the training data to construct the model ➤ connect the model and the scoring data to the Apply Model operator. That's it! You can now execute the process and the result is obtained. Yes, very straightforward. This is shown in Figure 1-1. Be aware that there is no model validation in this simple process.

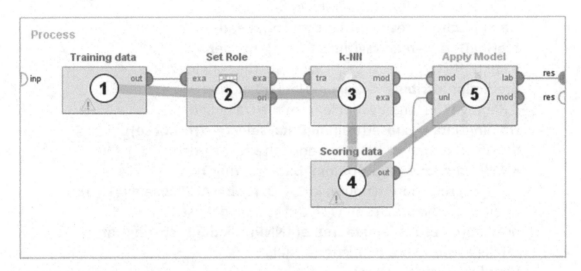

Figure 1-1. K-NN model in RapidMiner

Applying K-NN method is very simple in R, too. After loading the library "class", read the training data and scoring data, make use of the K-NN function, and by then we have finished our job: ready to view our result. This is demonstrated in Figure 1-2. Note that lines starting with "#" are comments.

3

```
# If the package 'kknn' has not been installed, install it.
# install.packages("kknn").
# Get a fresh start.
rm(list=ls())
# KNN is inside the library 'class'. Import the library.
library(class)
# Read the training and scoring datasets.
trainingData <- read.csv(file.choose(), header = T)
scoringData <- read.csv(file.choose(), header = T)
# Keep the target (response) values in the training data.
predictionAttribute <- trainingData$response
# Remove the target variable from the training data.
trainingData <- subset(trainingData, select=-c(response))
# Using the training data to predict the scoring data.
KNNPredictions <- knn(trainingData, scoringData,
        predictionAttribute, k=1, l=0, prob=FALSE, use.all=TRUE)
# Combine the prediction with the scoring data set.
predictionResults <- data.frame(KNNPredictions, scoringData)
# Take a look at the prediction results.
View(predictionResults)
```

Figure 1-2. *K-NN in R*

The knowledge you have gained from the preceding tasks is enough to just be able to apply the data mining method K-NN. But if you are trying to understand, step by step, why and how K-NN works, you will need a lot more information. Excel can offer you the opportunity to go through a step-by-step analysis process on a dataset during which you can develop a solid understanding of the K-NN algorithm. With this solid understanding, you can then be more proficient in using other powerful tools or programming languages. Most importantly, you will have a better understanding of the quality and value of your data mining results. You will see that in later chapters.

Of course, Excel is much more limited in data mining compared to Python, R, and RapidMiner. Excel can only work with data up to a smaller size limit. Meanwhile, some data mining techniques are too complicated to be practiced through Excel. Nonetheless, Excel provides us direct and visual understanding of the data mining mechanisms. In addition, Excel is naturally suitable for data preparation.

Today, because of the software tools and other packages, most effort in a data mining task is spent on understanding the task (including the business understanding and data understanding), preparing the data, and presenting the results. Less than 10% of the effort is spent on the modeling process. The process of preparing the data for modeling is called data engineering. Excel has an advantage on data engineering when the datasets are not too large because it can give us a visual representation of data engineering, which allows us to be more confident in our data preparation process.

As an experienced educator, I realize that students can better develop a deep understanding of data mining methods if these methods are also explained through step-by-step instructions in Excel. Studying through Excel unveils the mystery behind data mining or machine learning methods and makes students more confident in applying these methods.

Did I just mention machine learning? Yes, I did. Machine learning is another buzz phrase today. What is machine learning? What is the difference between data mining and machine learning? Moreover, what is artificial intelligence (AI) and what is the difference between AI and machine learning?

The purposes of machine learning and data mining are somewhat different. The purpose of machine learning is to study how computers can develop human-like learning ability by learning from data. The purpose of data mining is to find valuable patterns or knowledge from data. However, data mining makes use of machine learning methods to achieve its goals, that is, the methodologies of data mining and machine learning can be the same. Anyhow, it is not necessary to differentiate them, and I would suggest that we treat them the same at this moment.

The same scenario applies to AI and machine learning. Simply put, AI is computer software that simulates human brain functions, while machine learning trains computer algorithm through data to mimic human thinking ability. Because most AI makes use of machine learning to achieve its goals, machine learning is usually considered a subset of AI and therefore it is almost impossible to differentiate AI and machine learning either.

Why the Second Edition?

Practice is the key in learning, especially in learning technologies including Excel and data mining. Thus, the first reason for the second edition is to provide homework-like exercise(s) at the end of each chapter so that you can practice on your own the core skills explained in each chapter. Starting from this first chapter, you will find reinforcement exercises which may even be more challenging than whatever was explained step by step in the book. Of course standard answers or suggestions are provided to these exercises.

The second reason is that some chapters are updated based on feedbacks from some readers.

1. Chapter 3, K-Means Clustering: By introducing the function SHEET, the k-means clustering process is updated so that the process is even more automatic.

2. Chapter 6, Logistic Regression: As loss function is frequently used in machine learning, the concept and use of loss function is introduced in this chapter.

3. Chapter 9, Naive Bayes Classification: The Naïve Bayes' theorem is now explained in detail such that the theorem is deduced from the conditional probability. This gives readers a clear mathematical understanding about why and how the Naïve Bayes classification works.

4. Chapter 14, Text Mining: Because of the arrival of ChatGPT, I consider it is necessary to give a simple and quick introduction about the word vectorization (also called text embedding) technology which is widely used in large language models including ChatGPT.

The third reason is that some minor errors are fixed and some formulas are updated. For example, the function IFERROR is used more often to simplify a few formulas.

The last reason, probably the primary reason for the second edition, is that there are two new chapters.

1. One chapter is about exploratory data analysis (EDA), data cleaning, and feature selection. These critical concepts and skills are a must for studying data mining or machine learning. They are usually introduced before any data mining or machine learning models are explained. Though I take this opportunity to include them in this edition, I introduce them in Chapter 11 after several machine learning models. The reason is explained in Chapter 11. So, you will figure out why when you read Chapter 11.

2. Another new chapter (Chapter 8) is about hierarchical clustering which was once mentioned together with k-means clustering in the first edition of this book. This chapter presents two bright highlights. One is to implement a bottom-up hierarchical clustering algorithm in Excel; the second is to show how to draw a dendrogram in Excel without any add-in.

Prepare Some Excel Skills

There are quite some Excel skills to learn in this book. I will explain some of them in detail when we need to use them. However, there are several fundamental Excel skills and functions that we need to be familiar with before we start talking about data mining.

Formula

Formula is the most important feature of Excel. Writing a formula is like writing a programming statement. In Excel, a formula always starts with an equal sign ("=" without quotation marks).

Upon opening an Excel file, we are greeted with a table-like worksheet. Yes, every worksheet is a huge table. One reason why Excel is naturally suitable for data storage, analysis, and mining is because data are automatically arranged in a table format in Excel. Each cell in the big table has a name or so-called reference. By default, each column is labeled by a letter, while each row is labeled with a number. For example, the very first cell at the top-left corner is cell A1, that is, column A and row 1. The content in a cell, whatever it is, is represented by the cell reference.

Enter number 1 in cell A1. The value of cell A1 is 1 and A1 represents 1 at this moment.

Enter the formula "=A1*10" (without the double quotation marks and no space before "=") in cell B1 and hit the Enter key. Note that the formula starts with "=". Be aware that this is the only time a formula is presented inside a pair of double quotation marks in this book. From now on, all formulas are presented directly without quotation marks.

Enter the text "A1 * 10" in cell C1. Because the text does not start with "=", it is not a formula.

Our worksheet looks like Figure 1-3.

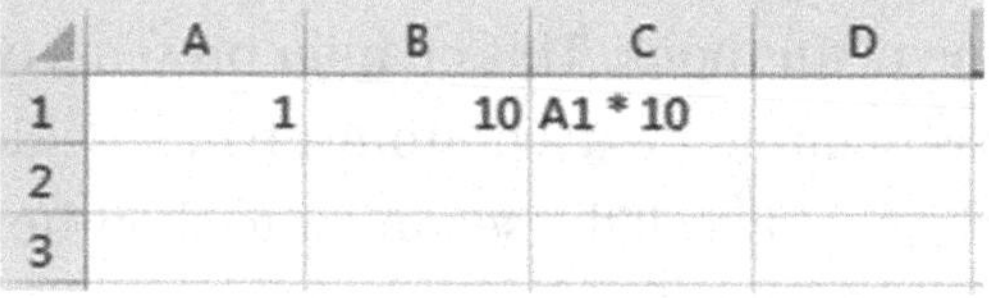

◢	A	B	C	D
1	1	10	A1 * 10	
2				
3				

Figure 1-3. *Excel formula*

Autofill or Copy

Autofill is another critical feature of Excel which makes Excel capable of working with a relatively large dataset. Autofill is also called "copy" by many people.

Let's learn autofill by the following experiment:

1. Enter 1 in cell A1.

2. Enter 2 in cell A2.

3. Select both cells A1 and A2.

4. Release the left mouse button.

5. Move the mouse cursor to the lower right corner of cell A2 until the cursor becomes a black cross (shown in Figure 1-4).

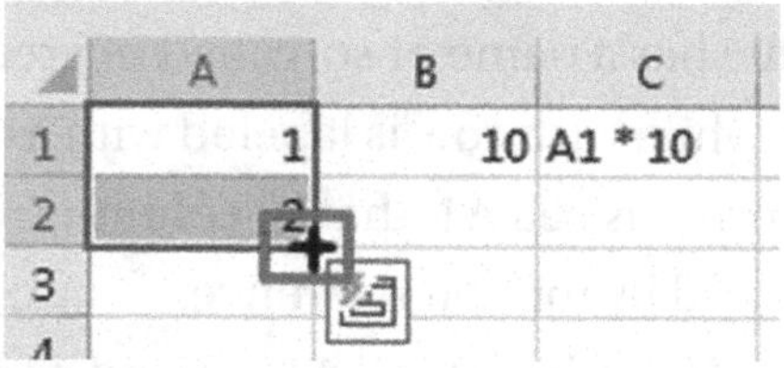

◢	A	B	C
1	1	10	A1 * 10
2	2		
3			

Figure 1-4. *Cursor becomes a black cross*

6. Press down the left mouse button and drag down to cell A6.

The cells A1:A6 are automatically filled with numbers 1, 2, 3, 4, 5, and 6. This process is called autofill. Some people call it "copy", too. But more precisely, this process is autofill.

Let's conduct another experiment:

1. Select cell B1 (make sure that B1 still has the formula =A1*10). Lift up the left mouse button.

2. Move the mouse cursor to the lower right corner of cell B1 until the cursor becomes a black cross.

3. Drag down the mouse cursor to cell B6. Our worksheet looks like Figure 1-5.

| B2 | ▼ | ⋮ | ✕ | ✓ | *fx* | =A2*10 |

◢	A	B	C	D	E
1	1	10	A1 * 10		
2	2	20			
3	3	30			
4	4	40			
5	5	50			
6	6	60			

Figure 1-5. *Autofill formula (relative reference)*

Note that in Figure 1-5, the formula in cell B1 is =A1*10, but the formula in cell B2 is automatically changed to =A2*10. When we autofill vertically down, Excel automatically increments the row index of all the cell references by 1 in a formula. Similarly, when we autofill horizontally to the right, Excel automatically increments the column index of all the cell references in a formula.

Click on cell B3; the formula inside must be =A3*10.

Note   Changing the number in A1 to 10 automatically changes the number in B1 to 100. By default, the whole Excel workbook will automatically recalculate if a cell value is changed. This is another important feature of Excel.

Absolute Reference

Assume that there is a ledger keeping the fixed interest rate and the amounts of loans lent to customers. Our job is to compute the interest each customer owes. Follow these instructions to complete the experiment:

1. Open a blank Excel worksheet. Enter texts "Loan", "Interest", and "Rate" in cells A1, B1, and D1. Leave C1 blank. Enter "5%" inside cell D2 (without quotation marks).

2. In cell A2, enter the formula `=RANDBETWEEN(1000,50000)`. This formula randomly generates an integer between 1000 and 50000 inclusively.

3. Autofill from A2 to A12.

4. In cell B2, enter the formula `=A2*D2`. This formula calculates how much interest A2 owes. Everything should be perfectly fine at this moment.

5. Let's try another quick autofill skill: select cell B2 ➤ move the mouse cursor to the lower right corner of cell B2 until the cursor becomes a black cross ➤ double-click. See Figure 1-6. This double-click action automatically fills in formulas from cell B3 to cell B12 (B2 already has a formula). Note: autofill by double-click works only with vertical downward autofill operations.

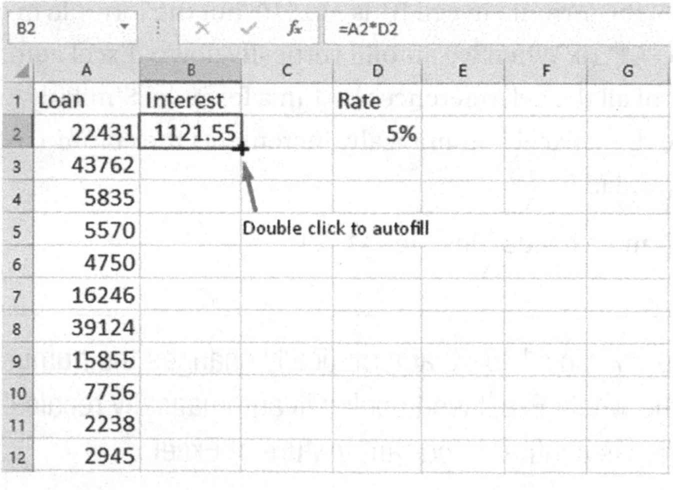

	A	B	C	D	E	F	G
1	Loan	Interest		Rate			
2	22431	1121.55		5%			
3	43762						
4	5835						
5	5570						
6	4750						
7	16246						
8	39124						
9	15855						
10	7756						
11	2238						
12	2945						

Figure 1-6. *Autofill by double-click*

Our worksheet now looks like Figure 1-7.

B3	▾	:	✕ ✓	fx	=A3 * D3

◢	A	B	C	D	E
1	Loan	Interest		Rate	
2	9767	488.35		5%	
3	41622	0			
4	6257	0			
5	48650	0			
6	46227	0			
7	4974	0			
8	38642	0			
9	37299	0			
10	41768	0			
11	8813	0			
12	12201	0			

Figure 1-7. Autofill failed on interest calculation

Except for cell B2, all other cells in column B do not obtain the correct result. The reason is because when we autofill from B2 to B12 vertically, the original formula in cell B2 (=A2*D2) changes: the row index of every cell reference in the formula is automatically incremented one by one. As shown in Figure 1-7, the formula in cell B3 is =A3*D3. We can imagine that the formula in cell B4 is =A4*D4.

In cell B3, we need the formula to be =A3*D2, that is, when we autofill from B2 to B3, we need the cell reference A2 to be changed to A3, but we need the cell reference D2 to stay the same.

As mentioned before, cell reference D2 has two parts: the column index D and the row index 2. For vertical autofill, the column index will never change but the row index will. To keep the row index unchanged during autofill, we need to place "$" before the row index. This means, the formula in cell B2 should be =A2*D$2. Using the symbol "$" to keep cell reference(s) in a formula unchanged or locked in autofill operations is called absolute reference.

With this corrected formula in cell B2, let's autofill from cell B2 to B12 again. We should get the correct result this time.

Be aware that in cell B2, the formula =A2*$D$2 works fine, too, for this specific task. When placing $ before the column index D, the column index won't change even when we are performing a horizontal autofill. However, be aware that we will run into cases where we must keep only part of a cell reference (either row index or column index only) unchanged in an autofill operation.

Paste Special and Paste Values

Continue with the worksheet on which we have been trying some interest calculation. Because they are generated by the function RANDBETWEEN, the numbers in column A keep changing in our worksheet and this is not good for demonstration. We would like to copy the numbers in cells A1:D12 to another location and remove the random number feature. We can accomplish this task by using the Paste Special feature.

Follow these instructions:

1. Select the cell area A1:D12. Under the main tab Home, click tool tab Copy. Let's copy the numbers to area A15:D26.

2. Click on cell A15; under the main tab Home, click Paste ➤ select Paste Special. This process is explained in Figure 1-8. Note if your operating system has Right Click enabled, you can right-click on cell A15 ➤ select Paste Special, too.

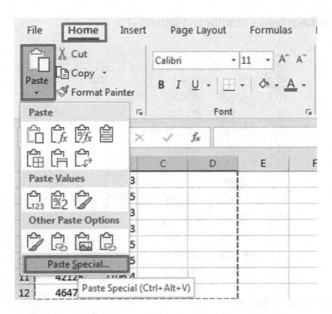

Figure 1-8. *Select Paste Special*

3. A small menu shows up. On this menu, choose Values under
    Paste as shown in Figure 1-9. Be aware of another available
    feature "Transpose" on this menu. It is a good idea to practice this
    Transpose feature as it is very useful for data preparation.

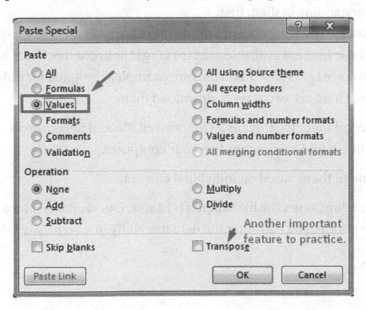

Figure 1-9. *Paste Values*

13

Part of our worksheet looks similar to Figure 1-10.

	A	B	C	D
15	Loan	Interest		Rate
16	40429	2021.45		0.05
17	41585	2079.25		
18	19405	970.25		
19	16896	844.8		
20	27130	1356.5		
21	49177	2458.85		
22	3855	192.75		
23	20231	1011.55		
24	46700	2335		
25	42610	2130.5		
26	30315	1515.75		

Figure 1-10. *After Paste Values*

IF Function Series

The IF statement is said to be the most used statement in programming, and this saying is also true in our learning data mining through Excel effort. As we are going to make use of the function IF and other IF-related functions very often, it is a good idea for us to get some basic understanding of them first.

This book comes with a number of sample Excel files. We are going to make use of them often. These files are available at https://github.com/hhohho/Learn-Data-Mining-through-Excel-2. These Excel files are examples designed for different data mining practices. There are two ways to download them:

1. Download them together as a compressed file and decompress them into a folder (directory) on your computer.

2. Download them based on individual chapters.

After downloading, open the file Chapter1-1a.xlsx. Our worksheet looks like Figure 1-11. If you want, you can enter the data manually in a blank worksheet, too.

▲	A	B	C	D
1	Name	Season	Sales	Commissions
2	Amie	1	$50,500	
3	Jessie	1	$100,200	
4	Jack	1	$86,000	
5	Jessie	2	$120,000	
6	Joshua	4	$87,000	
7	Amie	2	$98,000	
8	Jack	2	$75,000	
9	Amie	3	$110,000	
10	Amie	4	$88,000	

Figure 1-11. *The performances of some sales representatives*

Our first task is to compute the commissions for each sales representative. The commission is calculated in this way:

1. Any sale amount lower or equal to $50000 has no commission.

2. The sale amount more than $50000 but less than or equal to $100000 gains 10% commission.

3. The sale amount more than $100000 gains 20% commission.

Our formula to compute the commissions is based on the IF function. Enter the following formula in cell D2:

```
=IF(C2<=50000,0,IF(C2<=100000,(C2-50000)*10%,50000*0.1+(C2-100000)*20%))
```

IF function has the basic form: IF(boolean-expression, true-value, false-value). In our formula, if the sales in C2 are no more than 50000, 0 is returned. However, when C2 is more than 50000, a nested IF function is used here as we have two more conditions to assess.

In the inner IF function, if C2<=100000, the amount more than 50000 is multiplied by 0.1 (10%) and returned. Otherwise, C2 must be more than 100000, and therefore the calculated result of the expression 50000*0.1+(C2-100000)*20%) is returned.

Our worksheet looks exactly like Figure 1-12.

▲	A	B	C	D
1	Name	Season	Sales	Commissions
2	Amie	1	$50,500	50
3	Jessie	1	$100,200	5040
4	Jack	1	$86,000	3600
5	Jessie	2	$120,000	9000
6	Joshua	4	$87,000	3700
7	Amie	2	$98,000	4800
8	Jack	2	$75,000	2500
9	Amie	3	$110,000	7000
10	Amie	4	$88,000	3800

Figure 1-12. *Commissions calculated by using the IF function*

What if we want to count how many sales representatives have sales between 50000 and 100000 in season 1? To answer this question, we need to apply the function COUNTIFS. In Excel, there is another function named COUNTIF. COUNTIFS is capable of every task that COUNTIF is capable of, and there are many tasks for which we must use COUNTIFS instead of COUNTIF.

The explanation of COUNTIFS from Microsoft Office Support states "The COUNTIFS function applies criteria to cells across multiple ranges and counts the number of times all criteria are met." Its syntax is like COUNTIFS(criteria_range1, criteria_1, criteria_range2, criteria_2, ...). Based on this understanding, enter the following formula in cell E1:

```
=COUNTIFS(B2:B10,1,C2:C10,">50000",C2:C10,"<=100000")
```

This formula makes sure that

- In cell range B2:B10, the season is 1.

- And in cell range C2:C10, the sales are >50000. Note the criteria is placed inside a pair of double quotation marks.

- And in cell range C2:C10, the sales are <=100000.

The answer in cell E1 must be 2.

We want to examine the total sales and probably the averaged sales for every sales representative. For this purpose, we need to set up a table. Follow these instructions to set up the table:

1. Copy cells A1:A10 to F1:F10. Select cells F1:F10; follow instructions in Figure 1-13 to select Remove Duplicates (click main tab Data ➤ click the icon for Remove Duplicates).

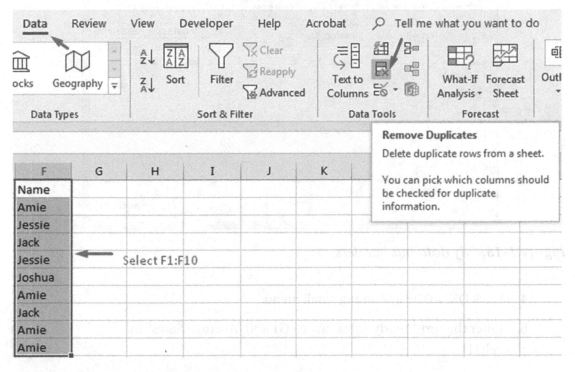

Figure 1-13. *Select the Remove Duplicates feature*

2. A small menu may show up like Figure 1-14. Make sure to select "Continue with the current selection". Click the button Remove Duplicates.

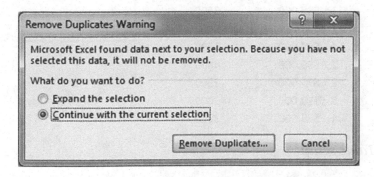

Figure 1-14. *Continue with the current selection*

3.   Check "My data has headers" as shown in Figure 1-15.

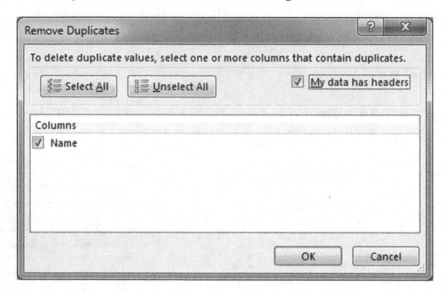

Figure 1-15. *My data has headers*

4.   Click OK on the upcoming small menu.

5.   Enter the text "Yearly Sales" in cell G1 and "Average Sales" in cell H1.

Our worksheet should look like Figure 1-16.

	A	B	C	D	E	F	G	H
1	Name	Season	Sales	Commissions	2	Name	Yearly Sales	Average Sales
2	Amie	1	$50,500	50		Amie		
3	Jessie	1	$100,200	5040		Jessie		
4	Jack	1	$86,000	3600		Jack		
5	Jessie	2	$120,000	9000		Joshua		
6	Joshua	4	$87,000	3700				
7	Amie	2	$98,000	4800				
8	Jack	2	$75,000	2500				
9	Amie	3	$110,000	7000				
10	Amie	4	$88,000	3800				

Figure 1-16. *Table setup*

For yearly sales computation, we need to use the function SUMIFS. Again, there is another function named SUMIF. The syntaxes of these two functions are different and could cause confusion. As SUMIFS can accomplish anything SUMIF can do, let's stick to SUMIFS for simplicity.

1. SUMIFS has the syntax SUMIFS(sum_range, criteria_range1, criteria1, criteria_range2, criteria2, …). Based on this understanding, enter the following formula in cell G2:

    =SUMIFS(C$2:C$10,A$2:A$10,F2)

    The preceding formula works this way: in the array A2:A10, cells A2, A7, A9, and A10 equal F2 ("Amie"); thus, their corresponding cells in the array C2:C10 are summed, that is, cells C2, C7, C9, and C10 are summed. Observe the use of absolute cell references in the formula because we are going to autofill from cell G2 to G5.

2. Autofill from cell G2 to cell G5.

3. Enter the following formula in cell H2 (here you can use either AVERAGEIF or AVERAGEIFS, they have similar syntax):

    =AVERAGEIFS(C$2:C$10,A$2:A$10,F2)

4. Autofill from cell H2 to cell H5.

Our worksheet looks exactly like Figure 1-17. The complete result is available in Chapter1-1b.xlsx.

	A	B	C	D	E	F	G	H
1	Name	Season	Sales	Commissions		2 Name	Yearly Sales	Average Sales
2	Amie	1	$50,500	50		Amie	346500	86625
3	Jessie	1	$100,200	5040		Jessie	220200	110100
4	Jack	1	$86,000	3600		Jack	161000	80500
5	Jessie	2	$120,000	9000		Joshua	87000	87000
6	Joshua	4	$87,000	3700				
7	Amie	2	$98,000	4800				
8	Jack	2	$75,000	2500				
9	Amie	3	$110,000	7000				
10	Amie	4	$88,000	3800				

Figure 1-17. *The use of functions SUMIFS and AVERAGEIFS*

Beware that since Office 2019 or Microsoft 365, Excel adds two more functions into the IF-function family: MAXIFS and MINIFS. Their syntaxes are very much similar to those of COUNTIFS, SUMIFS, and AVERAGEIFS.

Another two Excel features used often in data preparation are data import and export. To export data in Excel is fairly simple. For example, click File ➤ Save As ➤ select CSV to export the current worksheet into a file in CSV format. Note, CSV format is probably the most used file format for data science. Here, CSV represents "comma-separated values."

Today, Excel is capable of importing data from different resources. Click the main tab Data ➤ click Get Data; a list of data sources is presented including From File, From Databases, etc. For general purpose, importing data from files might be the most common. Be aware that Excel can only import text files properly. It cannot properly import files with special formatting such as Microsoft Word files.

Different Excel versions have different user interfaces on data importing. I won't illustrate how to use this feature in this book because all needed data for this book are well prepared in those Excel files which can be downloaded from GitHub at `https://github.com/hhohho/Learn-Data-Mining-through-Excel-2`.

Before I end this chapter, I would like to point out that this book is for Office 2016 and after. Thus, some new and more powerful functions of Office 365 are not introduced in this book because they are not available in Office 2016.

Reinforcement Exercises

If you would like to practice some of the skills mentioned in this chapter, please open the Chapter1-HW.xlsx file to practice. Should you run into difficulties, please take a look at the solutions inside Chapter1-HW-withAnswers.xlsx.

Review Points

This is our first chapter. Please review the following topics and skills:

1. The advantages of learning data mining through Excel examples

2. Excel formula

3. Excel skill autofill

4. Absolute and relative references

5. Remove Duplicates and Paste Special

6. Excel functions RANDBETWEEN, IF, COUNTIFS, SUMIFS, and AVERAGEIFS

## CHAPTER 2

Linear Regression

Please download the sample Excel files from `https://github.com/hhohho/Learn-Data-Mining-through-Excel-2` for this chapter's exercises.

General Understanding

Linear regression is a predictive model in which training data is employed to construct a linear model to make predictions on the scoring data. When we talk about a linear model, we mean that the relationship between the target (dependent variable) and the attribute(s) (independent variables) is linear. It is a convention to use the terminologies "independent variable" and "dependent variable" in regression analysis. Therefore, in this chapter, we will replace attribute with independent variable and substitute dependent variable for target.

There might be one or more independent variables in linear regression analysis. When there is only one independent variable, the linear model is expressed by the commonly seen linear function $y = mx + b$, where y is the dependent variable, m is the slope of the line, and b is the y-intercept. In most cases, there is more than one independent variable; therefore, the linear model is represented as $y = m_1x_1 + m_2x_2 + \ldots + m_nx_n + b$, where there are n independent variables and m_i is the coefficient associated with a specific independent variable x_i, for $i = 1 \ldots n$. To construct such a linear model, we need to find the values of m_i and b, based on the known y and x_i values in the training dataset.

Let's start a scenario to learn what linear regression can do. In a southern beach town, Tommy, the manager of a supermarket is thinking about predicting ice cream sales based on the weather forecast. He has collected some data that relate weekly averaged daily high temperatures to ice cream sales during a summer season. The training data are presented in Table 2-1.

© Hong Zhou 2023
H. Zhou, *Learn Data Mining Through Excel*, https://doi.org/10.1007/978-1-4842-9771-1_2

Table 2-1. *Ice cream sale vs. temperature*

Temperature (F)	Ice Cream Sale (Thousands of Dollars)
91	89.8
87	90.2
86	81.1
88	83.0
92.8	90.9
95.2	119.0
93.3	94.9
97.7	132.4

Enter the preceding data into an Excel worksheet. Observe that in this scenario, the Temperature is the only independent variable, and the Ice Cream Sale is the dependent variable. Based on the given data, Tommy draws a scatter plot. The chart looks like Figure 2-1.

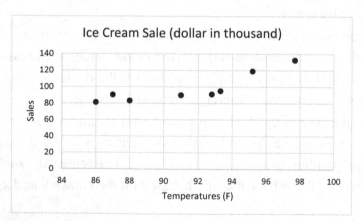

Figure 2-1. *Temperatures vs. ice cream sales*

Based on the chart in Figure 2-1, if the average of the daily high temperatures of next week is predicted to be 88.8 Fahrenheit degrees, Tommy finds it difficult to predict the sales since there is no dot directly matching to the temperature 88.8. Right-click on a dot; a small menu should show up. On the menu, click Add Trendline as shown in Figure 2-2.

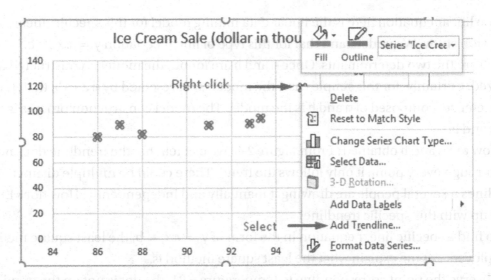

Figure 2-2. *Add Trendline*

A trendline shows up in the chart as illustrated by Figure 2-3. Be aware that by default, the line in Figure 2-3 is not a solid red line and the equation won't show up. It would be a good exercise to figure out how to format the trendline to be a solid red line and how to display the equation on the chart. To display the equation on the chart, right-click on the trendline in the chart ➤ select Format Trendline... ➤ check "Display Equation on chart".

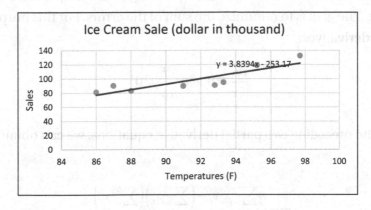

Figure 2-3. *Using trendline to predict*

With this linear line, Tommy can then estimate that when the temperature = 88.8, the sales would be about $89k. Using the equation $y = 3.8394x - 253.17$, where $m = 3.8394$ and $b = -253.17$, Tommy can predict his ice cream sales more precisely:

```
3.8394 × 88.8 - 253.17 = $87.8k.
```

The linear equation here is the linear data mining model for this specific linear regression study. We understand that for this type of linear equation y = mx + b, m and b are the two determinants. Once m and b are found, the model is constructed and finalized. Certainly, for this simple case, the model is represented by m and b, that is, the parameter set composed of m and b is the model. The model construction process is to find m and b.

How are m and b obtained? From Figure 2-3, we can tell that the trendline does not pass through every point, it only follows the trend. There could be multiple distinct trendlines if several people are drawing it manually and independently. How does Excel come up with this specific trendline?

To find a specific linear equation in the form of y = mx + b, the least square method is employed. Let me explain what the least square method is.

Observe the point corresponding to temperature = 91, the dot is not on the trendline. This indicates that there is an error (we can understand it as difference) between the trendline (the predicted value) and the actual data. In fact, all the data points in Figure 2-3 have small errors. The sum of the squares of the errors can be represented as

$$E = \sum_{i=1}^{n} \left(y_i - \left(mx_i + b \right) \right)^2 \tag{2-1}$$

where y_i is the actual ice cream sale corresponding to the temperature value x_i in the training dataset. The goal is to minimize the sum of the errors. For this purpose, we can take the partial derivatives:

$$\frac{\partial E}{\partial x} = 0 \text{ and } \frac{\partial E}{\partial y} = 0 \tag{2-2}$$

By solving the preceding two partial derivative equations, we can obtain the values of m and b as

$$m = \frac{n \sum_{i=1}^{n} x_i y_i - \left(\sum_{i=1}^{n} x_i \right)\left(\sum_{i=1}^{n} y_i \right)}{n \left(\sum_{i=1}^{n} x_i^2 \right) - \left(\sum_{i=1}^{n} x_i \right)^2} \tag{2-3}$$

$$b = \frac{\left(\sum_{i=1}^{n} x_i^2\right)\left(\sum_{i=1}^{n} y_i\right) - \left(\sum_{i=1}^{n} x_i\right)\left(\sum_{i=1}^{n} x_i y_i\right)}{n\left(\sum_{i=1}^{n} x_i^2\right) - \left(\sum_{i=1}^{n} x_i\right)^2} \tag{2-4}$$

Learn Linear Regression Through Excel

In Equations (2-3) and (2-4), x_i is the known averaged daily high temperature and y_i is the actual ice cream sale corresponding to x_i. They are from the training dataset shown in Table 2-1.

Open the file Chapter2-1a.xlsx. Enter the texts and formulas in columns C and D as shown in Figure 2-4 to compute the values of m and b. The formulas in Figure 2-4 are based on Equations (2-3) and (2-4). Observe that in Figure 2-4, temperature is renamed as x and ice cream sale is renamed as y so that the setup follows Equations (2-3) and (2-4) more closely.

◢	A	B	C	D
1	x	y		
2	91	89.8		
3	87	90.2	SUM(X)	=SUM(A:A)
4	86	81.1	SUM(Y)	=SUM(B:B)
5	88	83	SUM(XY)	=SUMPRODUCT(A:A, B:B)
6	92.8	90.9	SUM(X^2)	=SUMPRODUCT(A:A,A:A)
7	95.2	119	n	=COUNT(A:A)
8	93.3	94.9	m	=(D7*D5-D3*D4)/(D7*D6-D3^2)
9	97.7	132.4	b	=(D6*D4-D3*D5)/(D7*D6-D3^2)

Figure 2-4. *Computing m and b via least square approximation*

In Figure 2-4, A:A and B:B are used to represent all cells in column A and column B, respectively. Thus, =SUM(A:A) and =SUM(B:B) will sum up all cells in column A and column B, respectively. Also, an Excel function SUMPRODUCT is used to simplify the computation between two arrays. An important understanding of SUMPRODUCT is that the two arrays must be of the same type and length. Here, the same type means that the two arrays can be both a column and a row. With both m and b computed, Tommy can accomplish his predictions as shown in Figure 2-5 (enter the formula =D$8*C12 + D$9 inside cell D12 and autofill from D12 to D14).

◢	A	B	C	D	E
1	x	y			
2	91	89.8			
3	87	90.2	SUM(X)	731	
4	86	81.1	SUM(Y)	781.3	
5	88	83	SUM(XY)	71851.77	
6	92.8	90.9	SUM(X^2)	66915.06	
7	95.2	119	n	8	
8	93.3	94.9	m	3.83943386	
9	97.7	132.4	b	-253.165769	
10					
11			**Scoring Data**	**Prediction**	
12			88.8	87.8	
13			96.9	118.9	
14			94.7	110.4	
15					

Figure 2-5. *Prediction based on a linear regression model*

A much more efficient way to compute m and b is to make use of the two functions SLOPE and INTERCEPT for this specific example. Follow these instructions to practice the use of the two functions:

1. Input the x and y values as shown Figure 2-6. Note: input x and y values only. Columns C and D are empty.

2. Enter m in cell C2 and b in cell D2.

3. Enter the formula =SLOPE(B2:B9, A2:A9) in cell C3.

◢	A	B	C	D	E
1	X	Y			
2	91	89.8			
3	87	90.2			
4	86	81.1			
5	88	83			
6	92.8	90.9			
7	95.2	119			
8	93.3	94.9			
9	97.7	132.4			
10					

Figure 2-6. *Prepare data to use functions SLOPE and INTERCEPT*

4. Enter the formula =INTERCEPT(B2:B9, A2:A9) in cell D3.

The result should look like Figure 2-7.

◢	A	B	C	D	E
1	x	y			
2	91	89.8	m	b	
3	87	90.2	3.8394	-253.2	
4	86	81.1			
5	88	83			
6	92.8	90.9			
7	95.2	119			
8	93.3	94.9			
9	97.7	132.4			

Figure 2-7. *The result obtained by the SLOPE and INTERCEPT functions*

Another way to obtain the values of m and b is through the Data Analysis tool of Excel. This is a popular approach to take by many people, though I dislike it. The disadvantage of using Data Analysis tool for regression is that once the coefficients and intercept value are obtained, they won't update along with the changes in the training data.

By default, our Excel does not have Data Analysis tool available. To make it available, we must enable the built-in add-in Analysis ToolPak in our Excel. I will show you how to enable Analysis ToolPak and Solver in our Excel later when we need to use Solver. As of right now, we do not need to use the Data Analysis tool.

The results of the preceding learning processes are stored in the file Chapter2-1b.xlsx.

Learn Multiple Linear Regression Through Excel

Tommy has successfully predicted his ice cream sales, but he wants to improve his prediction even more. After some studies, he collected additional information such as the number of tourists (obtained from hotels) and the number of sunny days in a week. His updated training dataset is presented in Table 2-2.

Table 2-2. Ice cream sale vs. temperature, tourists, and sunny days

Temperature (F)	Tourists	Sunny Days	Ice Cream Sale (dollar in thousand)
91	998	4	89.8
87	1256	7	90.2
86	791	6	81.1
88	705	5	83
92.8	1089	3	90.9
95.2	1135	6	119
93.3	1076	4	94.9
97.7	1198	7	132.4

There are three independent variables in Table 2-2: temperature, tourists, and sunny days. When there is more than one independent variable, the linear regression is specifically called multiple linear regression. Let's denote temperature as x_1, tourists as x_2, sunny days as x_3, the y-intercept as b, and the sales as y. Our multiple linear regression equation is $y = m_1x_1 + m_2x_2 + m_3x_3 + b$. For such multiple linear regression, we need to make use of LINEST function to obtain the coefficients m_i and b. Follow these instructions to practice multiple linear regression through Excel:

1. Enter the data of Table 2-2 into cells A1:D9 in a new Excel worksheet, or open the file Chapter2-2a.xlsx.

The model construction process is to find the optimal values of m_1, m_2, m_3, and b. We are going to make use of the array function LINEST. Be advised that LINEST returns an array of values. The returned values are in the order of m_n, m_{n-1}, ..., m_1, and b, the reversed order to the way in which the formula is expressed. This necessitates the use of the INDEX function to fetch individual values from the returned array. The INDEX function is designed to work with a matrix or table. It requires a row index and a column index to locate an element in a table. As LINEST returns an array which has only one row, the row index is always 1, while the column indexes are 1, 2, 3, and 4, respectively.

2. Enter the texts "Sunny Days", "Tourists", "Temperature", and "Y-intercept" in cells A11:A14.

3. Enter numbers 1, 2, 3, and 4 in cells B11:B14. Our worksheet looks like Figure 2-8. I will explain why we need to enter these numbers soon.

	A	B	C	D
1	Temperature (F)	Tourists	Sunny Days	Ice Cream Sale (dollar in thousand)
2	91	998	4	89.8
3	87	1256	7	90.2
4	86	791	6	81.1
5	88	705	5	83
6	92.8	1089	3	90.9
7	95.2	1135	6	119
8	93.3	1076	4	94.9
9	97.7	1198	7	132.4
10				
11	Sunny Days	1		
12	Tourists	2		
13	Temperature	3		
14	Y-intercept	4		

Figure 2-8. *Set up the table to find coefficients and y-intercept*

4. Enter the following formula in cell C11 and hit Enter:

```
=INDEX(LINEST(D$2:D$9,A$2:C$9,TRUE,TRUE),1,B11)
```

In this formula, the first input argument in the function LINEST is D2:D9, the dependent variable values. The second input argument is A2:C9 which contains values for x_1, x_2, and x_3. The returned array from LINEST is fed to the function INDEX as an input. Since B11 = 1, the function INDEX fetches the first element in the returned array, which is m_3. Note again, LINEST returns an array that arranges the coefficients in the order of m_n, m_{n-1}, m_{n-2}, ..., m_1, b. In this specific example, the data in the returned array are m_3, m_2, m_1, and b for independent variables Sunny Days, Tourists, Temperature, and Y-intercept.

Refer to Figure 2-9.

	A	B	C	D
1	Temperature (F)	Tourists	Sunny Days	Ice Cream Sale (dollar in thousand)
2	91	998	4	89.8
3	87	1256	7	90.2
4	86	791	6	81.1
5	88	705	5	83
6	92.8	1089	3	90.9
7	95.2	1135	6	119
8	93.3	1076	4	94.9
9	97.7	1198	7	132.4
10				
11	Sunny Days		1	=INDEX(LINEST(D$2:D$9,A$2:C$9, TRUE, TRUE), 1, B11)
12	Tourists		2	
13	Temperature		3	
14	Y-intercept		4	

Figure 2-9. *Using INDEX and LINEST functions to obtain coefficients*

5. Autofill from cell C11 to cell C14. Pay attention to the use of B11, B12, B13, and B14.

When we autofill from cell C11 to C14, the preceding formula automatically becomes =INDEX(LINEST(D$2:D$9,A$2:C$9,TRUE, TRUE),1,B12) in cell C12. Because B12 = 2, the formula in C12 correctly fetches the coefficient for Tourists (m_2). The same logic applies to the formulas in C13 and C14. By pre-entering numbers in cells B11:B14, once we have the initial formula entered correctly, we can then obtain all needed values by autofill.

This is a common technique. We certainly do not want to enter the formula multiple times. Note, the function LINEST is repeatedly called for every coefficient.

6. Our last step is to examine how well the new model is. Enter the text "Predicted" in cell E1.

7. Enter the following formula in cell E2 and then autofill to cell E9:

=A2*C$13+B2*C$12+C2*C$11+C$14

As mentioned before, the parameter set, that is, the coefficients and the Y-intercept, represent the linear regression model. We want to examine how good the model is; therefore, we need to generate the predicted sales in column E.

8. To examine the quality of the preceding linear regression model, we can further compute the errors based on Equation (2-1). Enter the text "Error" in cell F1, and the formula =POWER(D2-E2,2) in cell F2, then autofill from cell F2 to F9.

9. Enter the text "Sum of Errors" in cell D11 and the formula =SUM(F2:F9) in cell E11. Our worksheet looks like Figure 2-10.

	A	B	C	D	E	F
1	Temperature	Tourists	Sunny Days	Ice Cream Sale (dollar in thousand)	Predicted	Error
2	91	998	4	89.8	88.7706	1.059665
3	87	1256	7	90.2	90.390726	0.036376
4	86	791	6	81.1	81.085358	0.000214
5	88	705	5	83	83.195593	0.038256
6	92.8	1089	3	90.9	89.847272	1.108235
7	95.2	1135	6	119	117.19569	3.25555
8	93.3	1076	4	94.9	97.80896	8.462051
9	97.7	1198	7	132.4	133.0058	0.367
10						
11	Sunny Days	1	5.95647	Sum of Errors	14.327347	
12	Tourists	2	-0.0013			
13	Temperature	3	3.97541			
14	Y-intercept	4	-295.47			

Figure 2-10. *A model of multiple linear regression*

Recall that the coefficients obtained through the function LINEST guarantee that the error value in cell E11 is minimized. You can manually modify the coefficients to examine if you can obtain a smaller error value. I will leave this as an exercise for you to practice.

The coefficient for the independent variable "Tourists" is negative. This simply means when there are more tourists, the ice cream sales will be smaller. However, in this specific example, this coefficient is close to zero, indicating that the number of tourists does not have much impact on the ice cream sales.

The complete work can be found in the file Chapter2-2b.xlsx.

Reinforcement Exercises

Practice is the key to success. It is a good idea to open Chapter2-HW.xlsx to practice linear regression by yourself. Chapter2-HW-withAnswers.xlsx provides one way of solutions to reference.

Review Points

We have reached the end of Chapter 2. For this chapter, please review the following concepts and Excel skills:

1. Linear regression model

2. Least square method

3. Multiple linear regression

4. Functions SUM, COUNT, SLOPE, INTERCEPT, LINEST, SUMPRODUCT, INDEX, and POWER

CHAPTER 3

K-Means Clustering

Please download the sample Excel files from https://github.com/hhohho/Learn-Data-Mining-through-Excel-2. Double-click Chapter3-1a.xlsx to open it.

General Understanding

Unlike linear regression which is a predictive method, clustering is a classification method that categorizes subjects (data points) into different groups (clusters), each with a number of characteristic measurements. For example, a bank may want to categorize customers into different risk groups when considering lending loans to its customers. Such a categorizing process relies on the selected characteristics and the algorithm adopted. Pay attention to the terminologies: subject, data point, group, cluster, class, etc. In a table, a row is called a record. Since a record can be composed of multiple numerical values, it can also be called a data point. In addition, a record can contain characteristics that are used to describe somebody, so in some cases a record is also called a subject. In this book, data point, record, and subject all have the same meaning. Similarly, the three terms group, cluster, and class have the same meaning, too.

   Clustering is an unsupervised data mining method. It does not require a training dataset. The two most popular types of clustering methods are partitioning clustering and hierarchical clustering. K-means clustering, where k represents the desired number of clusters, is a type of partitioning clustering. In k-means clustering, each cluster is defined by the centroid (or mean) of the data points in the cluster. For example, suppose a cluster has three data points expressed as three vectors: (1, 2, 3, 4, 5), (2, 3, 4, 5, 6), and (5, 4, 3, 2, 1). The centroid of this cluster is then ((1+2+5)/3, (2+3+4)/3, (3+4+3)/3, (4+5+2)/3, (5+6+1)/3), that is, (2.7, 3, 3.3, 3.7, 4). Note that the k-means clustering method requires all data to be numerical.

© Hong Zhou 2023
H. Zhou, *Learn Data Mining Through Excel*, https://doi.org/10.1007/978-1-4842-9771-1_3

To start the k-means clustering process, the first task is to decide how many clusters are needed, that is, the value of k. We may need to try multiple k values to examine which k value serves our purpose the best. The second task is to select k data points as the initial centroids. We can randomly select the k centroids, or we can pick them based on the data distribution. The distances from each data point to every centroid are computed, and a data point is pushed into a cluster that is the closest to it than all other clusters are. Once every data point has been pushed into a cluster, the centroid of the points in a cluster is recomputed. The distances of each data point to every new centroid are recomputed, too, followed by categorizing points into clusters based on the shortest distance rule again. Such a process is repeated until all data points are assigned into the same clusters in two consecutive iterations, at which point the cluster centroids have stabilized and will remain the same thereafter. The process can also be stopped when the predefined maximum number of iterations has been reached.

There are different methods to compute the distances between any two data points, though Euclidean distance is commonly used. Given two data points (x1, x2, x3, x4) and (y1, y2, y3, y4), their Euclidean distance is computed as

$$\sqrt{\left(x1 - y1\right)^2 + \left(x2 - y2\right)^2 + \left(x3 - y3\right)^2 + \left(x4 - y4\right)^2} \qquad (3\text{-}1)$$

Learn K-Means Clustering Through Excel

Suppose there are 50 countries that we want to categorize into four clusters based on four attributes: murder rate, assault rate, urban population, and rape rate. All the data are stored in the Excel file Chapter3-1a.xlsx. Let's open this Excel file which has only one worksheet named 1. Yes, this worksheet is named "1". Column A stores the unique codes for each country. The data look like Figure 3-1.

◢	A	B	C	D	E	F	G	H	I	J	K
1											
2	Country	Murder	Assault	UrbanPop	Rape						
3	L1	12	193	51	21						
4	L2	9	222	39	42						
5	L3	6	226	72	30						
6	L4	8	177	50	17						
7	L5	7	235	86	39						
8	L6	5	170	78	37						
9	L7	0	35	75	7						
10	L8	4	198	70	15						
11	L9	13	259	74	30						
12	L10	15	136	52	24						
13	L11	5	32	82	17						
14	L12	0	54	47	12						
15	L13	10	218	79	20						
16	L14	7	26	65	19						

Figure 3-1. *Data about 50 countries*

Let's select cells G1:K1 and then click "Merge & Center". Cells G1 to K1 are merged into one cell. Enter "Mean" in this merged cell (its reference is G1). Enter other data inside cells G2:K6 as shown in Figure 3-2. Note that C1, C2, C3, and C4 are all uppercased. Be advised that there is another file named Chapter3-1a-preEntered.xlsx in which data in G1:K6 are pre-entered. You are welcome to start with Chapter3-1a-preEntered.xlsx.

⊿	B	C	D	E	F	G	H	I	J	K
1								Mean		
2	Murder	Assault	UrbanPop	Rape		Cluster	C1	C2	C3	C4
3	12	193	51	21		Murder	2.3	5.6	11	15
4	9	222	39	42		Assault	15	53	100	200
5	6	226	72	30		UrbanPop	9.9	20	50	80
6	8	177	50	17		Rape	10	10	25	40
7	7	235	86	39						
8	5	170	78	37						
9	0	35	75	7						
10	4	198	70	15						
11	13	259	74	30						
12	15	136	52	24						
13	5	32	82	17						
14	0	54	47	12						
15	10	218	79	20						

Figure 3-2. *Set up data in G1:K6*

C1, C2, C3, and C4 represent cluster 1, cluster 2, cluster 3, and cluster 4, respectively. The numbers in cells H3:K6 are selected after a quick peek at the data ranges of each attribute. We can randomly pick some numbers for them, too. For this demonstration, let's keep these numbers exactly. Numbers in cells H3:H6, I3:I6, J3:J6, and K3:K6 stand for the centroids of C1, C2, C3, and C4, respectively.

Select cells M1:P1 and click "Merge & Center". Cells M1 to P1 are merged into one cell in which we enter the text "Distances". We are going to store the Euclidean distances from each data point to the four centroids in columns M, N, O, and P. The setup of cells M1:Q2 is illustrated in Figure 3-3.

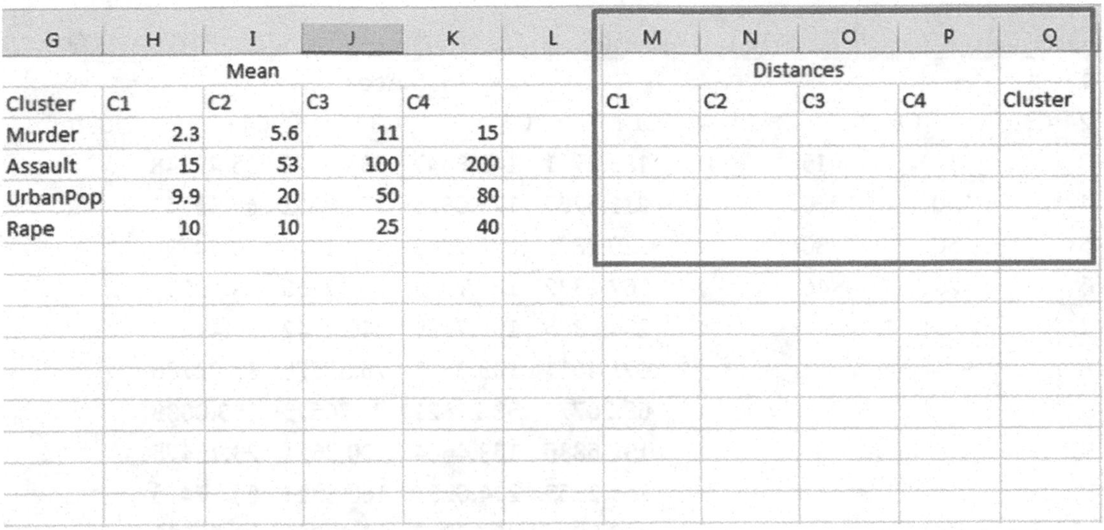

G	H	I	J	K	L	M	N	O	P	Q
		Mean						Distances		
Cluster	C1	C2	C3	C4		C1	C2	C3	C4	Cluster
Murder	2.3	5.6	11	15						
Assault	15	53	100	200						
UrbanPop	9.9	20	50	80						
Rape	10	10	25	40						

Figure 3-3. *Setup of the cells M1:Q2*

Cell M3 stores the distance from country L1 (12, 193, 51, 21) to centroid C1 (2.3, 15, 9.9, 10). The formula in M3 is based on Equation (3-1) and is presented as follows:

```
=SQRT(SUMXMY2($B3:$E3, H$3:H$6))
```

This formula computes the Euclidean distance between L1 and C1. Pay close attention to the use of absolute references in the formula. The use of absolute references here is for easy autofill. Function SQRT computes the square root of a value, and function SUMXMY2 computes the sum of squares of differences of corresponding values in two arrays (in this case, the two arrays are B3:E3 for L1 and H3:H6 for C1).

Continue the process by following these instructions:

1. Select cell M3 and autofill it horizontally to P3.

2. Select cells M3:P3; autofill them together to cells M52:P52. By far, the distances from each data point to every centroid have been computed. Part of our worksheet looks like Figure 3-4.

◢	J	K	L	M	N	O	P	Q
1				Distances				
2	C3	C4		C1	C2	C3	C4	Cluster
3	11	15		183.2711	143.9547	93.09672	35.49648	
4	100	200		211.5767	173.0825	123.6851	46.95743	
5	50	80		220.8871	181.7503	128.1015	30.34798	
6	25	40		167.1332	127.7919	77.47258	44.79955	
7				234.6365	195.7625	140.4742	36.41428	
8				171.4611	133.3505	76.57676	31.82766	
9				68.20777	58.21821	72.76675	169.0089	
10				192.6886	153.4684	100.7621	29.15476	
11				253.2969	214.0251	160.8913	60.17475	
12				129.5017	90.53927	36.29049	71.66589	

Figure 3-4. *Distances computed in worksheet 1*

3. Data in column Q specify the clusters to which each data point belongs based on the shortest distance rule. In cell Q3, enter the following formula:

```
=INDEX($M$2:$P$2,1,MATCH(MIN(M3:P3),M3:P3,0))
```

The syntax of function MATCH is MATCH(lookup_value, lookup_array, match_type). If match_type = 0, the function is looking for an exact match in the lookup_array for the lookup_value. If not found, an #N/A error (value not available error) is returned. If found, the function returns the relative position in the array. Note, the first position in Microsoft Office tools is always 1.

In the preceding formula, the MATCH function finds the relative column position of the minimum value of the array M3:P3 (which is 4 in this specific case). The column position is fed to the INDEX function to find the proper cluster name in the array M2:P2. This formula determines in which cluster the data point L1 is.

As there are only four clusters, a nested IF statement instead of the preceding formula suffices the purpose, too.

```
=IF(MIN(M3:P3)=M3,"C1",IF(MIN(M3:P3)=N3,"C2", IF(MIN(M3:P3)
=O3,"C3","C4")))
```

In the nested IF formula, if the minimum value is the same as M3, return "C1"; else if the minimum value equals N3, return "C2"; else if the minimum value equals O3, return "C3"; else returns "C4".

We can tell when there are too many clusters, INDEX and MATCH functions are a better programming approach than nested IF statements.

4. Autofill Q3 to Q52. By now, our data look like Figure 3-5.

J	K	L	M	N	O	P	Q
			Distances				
C3	C4		C1	C2	C3	C4	Cluster
11	15		183.2711	143.9547	93.09672	35.49648	C4
100	200		211.5767	173.0825	123.6851	46.95743	C4
50	80		220.8871	181.7503	128.1015	30.34798	C4
25	40		167.1332	127.7919	77.47258	44.79955	C4
			234.6365	195.7625	140.4742	36.41428	C4
			171.4611	133.3505	76.57676	31.82766	C4
			68.20777	58.21821	72.76675	169.0089	C2
			192.6886	153.4684	100.7621	29.15476	C4
			253.2969	214.0251	160.8913	60.17475	C4
			129.5017	90.53927	36.29049	71.66589	C3

Figure 3-5. *After first round of computation in worksheet 1*

5. We now need to start the second round of the clustering process. Make a copy of worksheet 1 and rename the new worksheet 2. Yes, the name of the new worksheet is number 2! Also, make sure worksheet 2 is right to worksheet 1. Their order is very important.

In worksheet 2, we need a formula in cell H3 to recalculate the mean of attribute Murder for all data points inside cluster 1. The needed data are inside the worksheet 1, however. A formula like the following can accomplish the job for us:

```
=AVERAGEIFS('1'!$B$3:'1'!$B$52, '1'!$Q$3:'1'!$Q$52,H$2)
```

Though this formula is inside worksheet 2, the data used are all in worksheet 1, and '1'! references worksheet 1 by its name which is "1" indeed. This formula makes use of the function AVERAGEIFS to compute the mean of attribute Murder for all data points inside cluster 1. Note, only the data points (column B only, too) whose cluster name matches cell H2 (C1) are averaged.

However, the preceding formula is not good enough!

First, this formula is computing an average. In the process of k-means clustering, it is possible that one cluster has no data points. In this case, the denominator will be zero and will result in a divided-by-zero error. Thus, we need to make use of IFERROR function to safeguard such a case.

Second and the most important reason is that, when we need to start another clustering iteration in another worksheet 3, we must go to cell H3 of worksheet 3 to change "1" to "2", because worksheet 3 must make use of data in worksheet 2. We do not want to manually revise our formulas again and again.

To answer this challenge, we need to make use of the functions SHEET, INDIRECT, and ADDRESS and operator &.

SHEET is a very interesting function which returns the sheet number of the referenced sheet. The leftmost worksheet in a workbook is numbered 1, the second leftmost worksheet is numbered 2, and so on. The syntax of SHEET is SHEET(reference), where the argument is usually a cell reference, for example, A1.

Function INDIRECT evaluates a text as a cell reference. Function ADDRESS converts two numbers into a cell reference. Function ADDRESS requires a row number and a column number. In Excel, column A is 1, B is 2, and so on. The operator & can concatenate texts (strings) together.

- In cell F1 of worksheet 2, enter the formula =SHEET(A1)-1. You will notice that the value in F1 is 1 which can be used to reference worksheet 1.

- In cells F3 to F6, enter numbers 2, 3, 4, and 5, respectively. These numbers can help to pick the right column when we autofill our formula horizontally. Again, in Excel, column B is column 2, column C is column 3, and so on. These numbers can be used to reference columns B, C, D, and E.

Part of our worksheet 2 looks like Figure 3-6.

	B	C	D	E		F	G	H	I	J	K
1						1			Mean		
2	Murder	Assault	UrbanPop	Rape			Cluster	C1	C2	C3	C4
3	12	193	51	21		2	Murder	2.3	5.6	11	15
4	9	222	39	42		3	Assault	15	53	100	200
5	6	226	72	30		4	UrbanPop	9.9	20	50	80
6	8	177	50	17		5	Rape	10	10	25	40
7	7	235	86	39							
8	5	170	78	37							
9	0	35	75	7							

Figure 3-6. *Set up worksheet 2 for formula automatic updates*

In cell H3, enter the following formula:

```
=IFERROR(AVERAGEIFS(INDIRECT($F$1 & "!" & ADDRESS(3,$F3,1))
: INDIRECT($F$1 & "!" & ADDRESS(52,$F3,1)),INDIRECT($F$1 &
"!$Q$3"):INDIRECT($F$1 & "!$Q$52"),H$2),0)
```

In this formula

- IFERROR makes sure when there is an error, the average is set to 0.

- ADDRESS(3,$F3,1) gives $B$3.

- INDIRECT($F$1 & "!" & ADDRESS(3,$F3,1)) → '1'!$B$3.

- INDIRECT($F$1 & "!" & ADDRESS(52,$F3,1)) → '1'!$B$52.

- INDIRECT($F$1 & "!" & ADDRESS(3,$F3,1)) : INDIRECT($F$1 & "!" & ADDRESS(52,$F3,1)) → '1'!$B$3:'1'!$B$52.

- INDIRECT($F$1 & "!$Q$3"):INDIRECT($F$1 & "!$Q$52") → '1'!$Q$3:'1'!$Q$52.

The preceding formula generates a final formula which is the same as

=IFERROR(AVERAGEIFS('1'!$B$3:'1'!$B$52, '1'!$Q$3:'1'!$Q$52,H$2),0)

The advantages of the preceding formula are as follows:

- We can autofill from H3 to K6.

- Together with the use of the SHEET function in cell F1, we can automate the referencing to worksheet 2 in worksheet 3 as long as we place worksheet 3 to the immediate right of worksheet 2.

6. Autofill from cell H3 to cell K6. Our worksheet can be compared to Figure 3-7.

F	G	H	I	J	K
1			Mean		
	Cluster	C1	C2	C3	C4
2	Murder	2	2.82	6.692308	9.789474
3	Assault	4.175	45.9	108.2308	215.3684
4	UrbanPop	52.125	55.5	64.84615	65.10526
5	Rape	10.875	12.7	19.69231	26.15789

Figure 3-7. *The centroids in worksheet 2*

7. By now, the distances of each data point to the four new centroids have been automatically computed, together with their belonging clusters. In cells R1 and R2, enter numbers 1 and 2, respectively. Select these two cells and autofill to cell R52. Cells R1:R52 now have numbers 1, 2, 3, ..., 52.

8. In cell S2, enter "Old cluster". Column S stores the belonging clusters of all the data points in the previous iteration (such data are in worksheet 1).

9. In cell T2, enter "difference". Column T books the cluster switches between the current iteration and the last iteration. Part of our worksheet looks like Figure 3-8.

M	N	O	P	Q	R	S	T
Distances					1		
C1	C2	C3	C4	Cluster	2	Old cluster	difference
189.3638	147.6883	86.06637	27.03322	C4	3		
220.5397	179.3882	118.8041	31.25793	C4	4		
223.569	181.7076	118.4377	13.77279	C4	5		
173.0506	131.3879	70.41714	42.27735	C4	6		
235.0398	193.3862	129.9646	31.5387	C4	7		
169.8793	128.4613	65.50482	48.63186	C4	8		
38.63239	23.22719	75.31087	181.9163	C2	9		
194.7015	152.8115	90.07965	21.99175	C4	10		
256.712	214.8413	151.5295	44.8096	C4	11		
133.1132	91.68589	31.99575	80.64056	C3	12		
41.39157	30.3101	78.20164	184.4349	C2	13		
50.14042	12.09555	57.995	163.2907	C2	14		
215.8487	173.9986	110.7278	15.42572	C4	15		

Figure 3-8. Ready to compare with worksheet 1

10. In cell S3, enter the following formula and autofill to S52:

    =INDIRECT($F$1 & "!Q" & R3)

    This formula references the value in cell Q3 of worksheet 1.

11. In cell T3, enter the following formula and autofill to T52:

    =IF(Q3=S3,0,1)

The preceding formula defines that for a given data point, if its last belonging cluster is different from the current belonging cluster, 1 is returned; otherwise, 0 is returned. Our goal is to reach a stable condition when column S has 0s only, that is, all data points stay in the same cluster as they were in the last iteration.

12. Select cells G10 and H10, merge them, and type "convergence" inside. In cell I10, enter the following formula:

=SUM(T3:T52)

As mentioned earlier, we want to get 0 inside the cell I10. Part of our worksheet 2 should look like Figure 3-9.

F	G	H	I	J	K	L	M	N	O	P	Q	R	S	T
1			Mean						Distances			1		
	Cluster	C1	C2	C3	C4		C1	C2	C3	C4	Cluster	2	Old cluster	Difference
2	Murder	2	2.82	6.69231	9.78947		189.364	147.688	86.0664	27.0332	C4	3	C4	0
3	Assault	4.175	45.9	108.231	215.368		220.54	179.388	118.804	31.2579	C4	4	C4	0
4	UrbanPop	52.13	55.5	64.8462	65.1053		223.569	181.708	118.438	13.7728	C4	5	C4	0
5	Rape	10.88	12.7	19.6923	26.1579		173.051	131.388	70.4171	42.2773	C4	6	C4	0
							235.04	193.386	129.965	31.5387	C4	7	C4	0
							169.879	128.461	65.5048	48.6319	C4	8	C4	0
							38.6324	23.2272	75.3109	181.916	C2	9	C2	0
	convergence		4				194.701	152.811	90.0796	21.9918	C4	10	C4	0
							256.712	214.841	151.529	44.8096	C4	11	C4	0
							133.113	91.6859	31.9957	80.6406	C3	12	C3	0
							41.3916	30.3101	78.2016	184.435	C2	13	C2	0
							50.1404	12.0956	57.995	163.291	C2	14	C2	0

Figure 3-9. *After the second round in worksheet 2*

13. As there are four data points that have switched clusters in the second iteration, we need continue the clustering process. Make a copy of worksheet 2 and give the new worksheet the name 3. Make sure that the worksheet 3 is immediately right to worksheet 2. We shall notice that worksheet 3 is automatically done and part of it looks like Figure 3-10.

F	G	H	I	J	K	L	M	N	O	P	Q	R	S	T
2			Mean						Distances			1		
	Cluster	C1	C2	C3	C4		C1	C2	C3	C4	Cluster	2	Old cluster	Difference
2	Murder	2	2.65455	7.66667	9.8125		189.364	146.244	73.5445	35.7077	C4	3	C4	0
3	Assault	4.175	47.3636	121.333	226.063		220.54	177.946	106.85	28.2629	C4	4	C4	0
4	UrbanPop	52.13	57.0909	66.8667	62.75		223.569	180.026	105.327	10.3761	C4	5	C4	0
5	Rape	10.88	13.7273	19.5333	27.25		173.051	129.981	58.2219	51.7498	C4	6	C4	0
							235.04	191.574	116.9	27.6842	C4	7	C4	0
							169.879	126.586	52.9584	59.1083	C3	8	C4	1
							38.6324	22.9325	87.9515	192.773	C2	9	C2	0
	convergence		4				194.701	151.2	76.9519	31.9987	C4	10	C4	0
							256.712	213.185	138.351	35.0594	C4	11	C4	0
							133.113	90.2234	22.58	90.9081	C3	12	C3	0
							41.3916	29.5418	90.6807	195.343	C2	13	C2	0
							50.1404	12.4859	71.0211	173.731	C2	14	C2	0

Figure 3-10. *K-means clustering after three rounds in worksheet 3*

14. Repeat the preceding clustering process: copy worksheet 3 to worksheet 4; make sure that the new worksheet is named 4 and is right to worksheet 3. Part of worksheet 4 looks like Figure 3-11.

F	G	H	I	J	K	L	M	N	O	P	Q	R	S	T
3			Mean						Distances			1		
	Cluster	C1	C2	C3	C4		C1	C2	C3	C4	Cluster	2	Old cluster	Difference
2	Murder	1.78889	3.34167	7.78571	10.133		186.873	139.068	64.9344	38.7852	C4	3	C4	0
3	Assault	6.71111	54.5	130.714	229.8		218.068	170.746	98.4446	28.5673	C4	4	C4	0
4	UrbanPop	51.2222	58.4167	68.8571	61.733		221.187	172.647	95.8334	12.1856	C4	5	C4	0
5	Rape	10.3333	15.75	20.4286	26.6		170.537	122.883	50.0975	54.9747	C3	6	C4	1
							232.754	184.106	107.308	27.9193	C4	7	C4	0
							167.636	119.072	43.6959	63.0488	C3	8	C3	0
							37.1477	27.2578	97.1592	196.494	C2	9	C2	0
	convergence		2				192.278	143.97	67.6201	35.3802	C4	10	C4	0
							254.325	205.814	128.851	31.9826	C4	11	C4	0
							130.681	82.9904	19.414	94.4649	C3	12	C3	0
							40.5161	32.6609	99.6833	199.133	C2	13	C2	0
							47.5399	12.4828	80.5883	177.309	C2	14	C2	0

Figure 3-11. *K-means clustering after four rounds in worksheet 4*

15. Repeat the clustering process again: copy worksheet 4 to worksheet 5; make sure that the new worksheet is named 5 and is right to worksheet 4. The convergence becomes 0 as shown in Figure 3-12.

F	G	H	I	J	K	L	M	N	O	P	Q	R	S	T
4			Mean						Distances			1		
	Cluster	C1	C2	C3	C4		C1	C2	C3	C4	Cluster	2	Old cluster	Difference
2	Murder	2.31	3.00909	7.8	10.2857		184.881	136.489	61.6318	42.6894	C4	3	C4	0
3	Assault	8.64	57.0909	134	233.571		216.104	168.195	95.2569	30.1276	C4	4	C4	0
4	UrbanPop	52.6	57.8182	67.6	62.5714		219.063	170.153	92.8412	13.1133	C4	5	C4	0
5	Rape	11.2	15.4545	20.2	27.2857		168.576	120.277	46.7577	58.9015	C3	6	C3	0
							230.541	181.704	104.566	26.4378	C4	7	C4	0
							165.394	116.722	41.3362	66.3451	C3	8	C3	0
							34.9226	29.3897	100.256	200.256	C2	9	C2	0
	convergence		0				190.203	141.439	64.5669	38.8711	C4	10	C4	0
							252.202	203.324	125.853	28.1418	C4	11	C4	0
							128.631	80.4817	17.7336	98.3105	C3	12	C3	0
							38.091	34.938	102.901	202.836	C2	13	C2	0
							45.7697	12.1481	83.1894	181.184	C2	14	C2	0

Figure 3-12. *K-means clustering after five rounds in worksheet 5*

16. Once the convergence has reached 0, no matter how many more iterations we continue, no data will change again. However, it is necessary to confirm it by creating another worksheet 6 to repeat the preceding process.

Note, it is very important to arrange the six worksheets in order; otherwise, you won't get the correct result. Take a look at Figure 3-13.

	B	C	D	E	F	G	H	I	J
1					5			Mean	
2	Murder	Assault	UrbanPop	Rape		Cluster	C1	C2	C3
3	12	193	51	21	2	Murder	2.31	3.009091	7.8
4	9	222	39	42	3	Assault	8.64	57.09091	133.8
5	6	226	72	30	4	UrbanPop	52.6	57.81818	67.6
6	8	177	50	17	5	Rape	11.2	15.45455	20.2
7	7	235	86	39					
8	5	170	78	37					
9	0	35	75	7					
10	4	198	70	15		convergence		0	
11	13	259	74	30					
12	15	136	52	24					

1 2 3 4 5 **6** ⊕

Figure 3-13. *The six worksheets must be placed in order*

By now, all data points have been successfully clustered.

From the preceding clustering process, we can tell that clustering in Excel is not difficult. Instead, it allows us to go through every single step and therefore develop a solid understanding of how k-means clustering works. I would like to challenge you to examine the numbers inside the four clusters, respectively. I am sure that you will find out some patterns. When you do so, remember that clustering is to group similar records together.

The complete result is in Chapter3-1b.xlsx (it also contains my observation on patterns).

Reinforcement Exercises

The file Chapter3-HW.xlsx provides an exercise opportunity for you. In this exercise, you need go over 15 worksheets to reach convergence. Should you find the exercise difficult, please refer to Chapter3-HW-withAnswers.xlsx.

The file Chapter3-HW-withAnswers-NameManager.xlsx makes use of Name Manager, an Excel built-in feature, to make the formulas simpler.

So far we employ the SHEET function to help automate the k-means clustering process. One critical requirement of the use of SHEET function is to place all worksheets in order. If we do not want to place all the worksheets in order, we can make use of the function CELL instead of SHEET. The file Chapter3-HW-withAnswers-Cell-Function.xlsx shows such a solution. Feel free to explore it.

Excel has some options to configure formula calculations. By default, the "Enable iterative calculation" option is disabled. If we enable iterative calculation in Excel, we can then complete the k-means clustering process in only two worksheets because the second worksheet can perform a loop-like procedure to reach convergence! Due to the space limit of this book, I won't explain how to achieve it. But please be advised that the clustering result obtained based on the preceding worksheet-by-worksheet procedure may be different from what can be acquired through this iterative calculation, though both results are correct, which indicates that there can be more than one optimal clustering solution.

Review Points

For Chapter 3, the following knowledge, skills, and functions are worth reviewing:

1. Centroid, Euclidean distance

2. Reference a worksheet by its name followed by the exclamation symbol "!"

3. Functions INDEX, IF, SQRT, SUMXMY2, and MATCH

4. Functions IFERROR, SHEET, INDIRECT, ADDRESS, and AVERAGEIFS

5. Operator &

6. Worksheet copying to start a new iteration

7. The clustering process

CHAPTER 4

Linear Discriminant Analysis

Please download the sample Excel files from `https://github.com/hhohho/Learn-Data-Mining-through-Excel-2` for this chapter's exercises.

General Understanding

Linear discriminant analysis (LDA) is similar to linear regression and k-means clustering, but different from both, too. K-means clustering is an unsupervised classification method; LDA is supervised since its classification is trained on known data, however. The predictive results are quantitatively numerical in linear regression, while they are qualitatively categorical in LDA.

LDA tries to find the best linear function that can separate data points into distinct categories or groups. The variances between group means should be as large as possible, while the variances inside each group should be as small as possible. Note that the variances are computed after every data point has been projected to a linear function $P = w_1x_1 + w_2x_2 + \ldots + w_nx_n + b$. The training dataset is used to find the best P function which is represented by the parameter set $(w_1, w_2 \ldots, w_n, b)$. Figure 4-1 demonstrates the concept in two dimensions.

© Hong Zhou 2023
H. Zhou, *Learn Data Mining Through Excel*, https://doi.org/10.1007/978-1-4842-9771-1_4

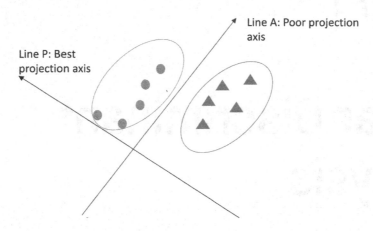

Figure 4-1. *LDA is to separate groups*

Given a set S of training data points which have already been sorted into different groups, the three steps of applying LDA can be simplified as follows:

Step 1: Sum the inter-group variances via the following expression:

$$Sg = \sum_{i}^{k} s_g = \sum_{i=1}^{k} (\bar{x}_i - \bar{x})^2 \tag{4-1}$$

where k is the number of groups, $\bar{x}$ is the mean of group means, and $\bar{x}_i$ is the mean of an individual group.

Step 2: Sum the within-group variances via the following expression:

$$S_w = \sum_{i}^{k} \sum_{j}^{n} (x_{ij} - \bar{x}_i)^2 \tag{4-2}$$

where n is the number of data points in a group.

Step 3: P is the best projection linear function such that Sg(P)/Sw(P) is maximized. As P is defined by $(w_1, w_2, \ldots, w_n, b)$, the model construction phase of LDA is to solve for $(w_1, w_2, \ldots, w_n, b)$. This is explained in Equation (4-3):

$$w_1 x_1 + w_2 x_2 + \ldots + w_n x_n + b = \max\left(Sg(P)/Sw(P)\right) \tag{4-3}$$

The least square method can be applied to solve for $(w_1, w_2, \ldots, w_n, b)$ first. We can then further optimize $(w_1, w_2, \ldots, w_n, b)$ by using Excel Solver.

Solver

Excel Solver is an optimization tool. It adjusts the values of decision variables to search for a maximum, minimum, or exact target value. Solver is a built-in add-in program that can be installed without any downloading. Follow these instructions to install Solver together with Data Analysis add-in:

1. Open a blank Excel file.

2. Click File ➤ Options. The link Options is at the bottom as shown in Figure 4-2.

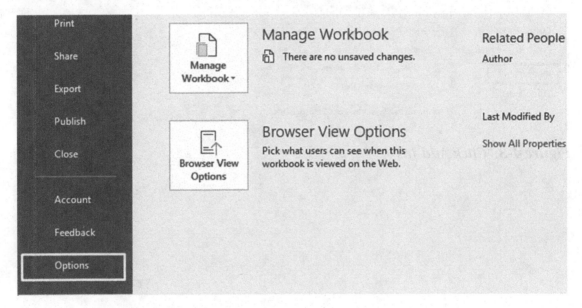

Figure 4-2.   *Click File ➤ Options*

3.   Click Add-ins as shown in Figure 4-3.

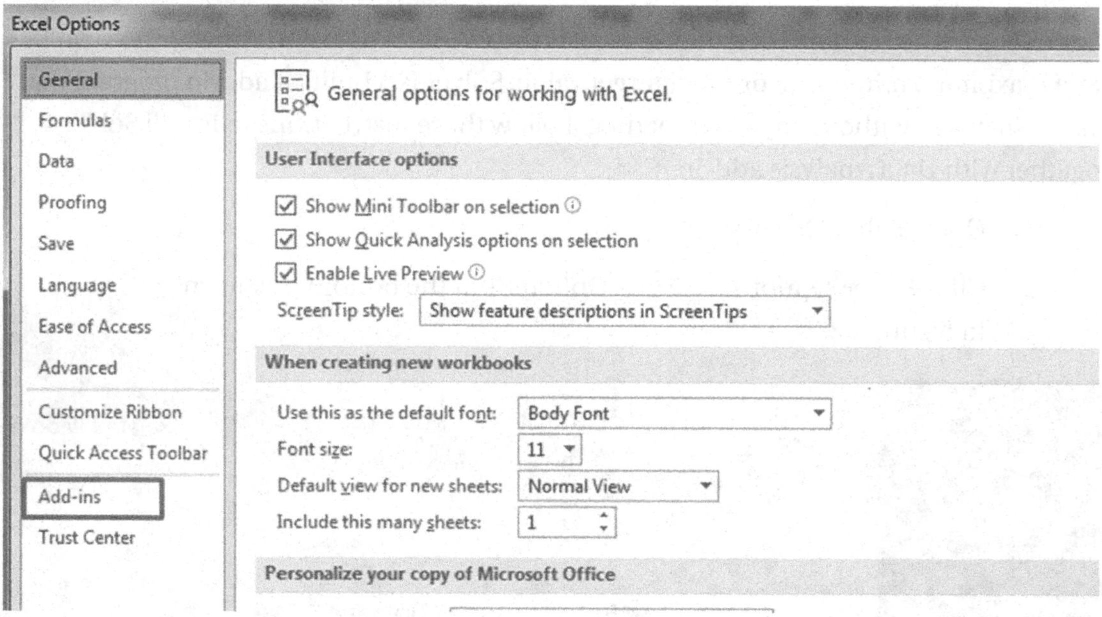

Figure 4-3. *Click Add-ins*

4. On the upcoming menu, select Solver Add-in, then click the
   button Go... as shown in Figure 4-4. Also, select Analysis ToolPak
   and click the button Go..., too.

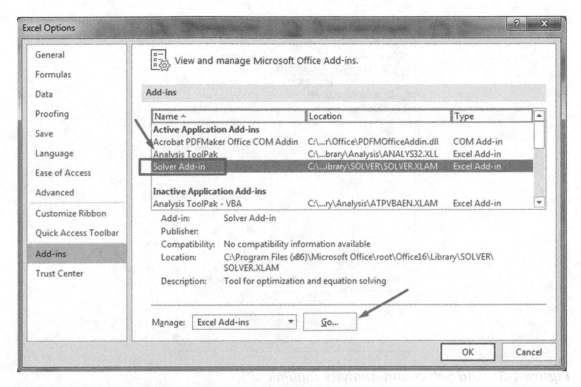

Figure 4-4. *Select Solver Add-in and click Go*

5. Click the button OK as shown in Figure 4-4. A small menu comes up as shown in Figure 4-5. Make sure that both Solver and Analysis ToolPak are checked. Click the button OK.

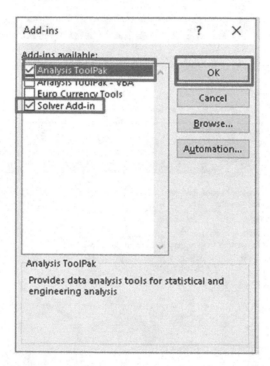

Figure 4-5. *Add Solver and Analysis ToolPak*

6. Click the main tab Data; both Data Analysis and Solver are available as shown in Figure 4-6.

Figure 4-6. *Solver is now available*

Learn LDA Through Excel

Chapter4-1a.xlsx contains the famous Iris dataset that is a perfect sample for LDA practice. In the Iris dataset, the four attributes x1, x2, x3, and x4 are some flower characteristics that are used to determine the type of iris: the target variable. This is the training dataset. Note that there are three distinct values for the target variable, that is, the target variable has three classes (categories). Follow these instructions:

1. Open Chapter4-1a.xlsx; enter "y" in cell F1 (without quotation marks).

2. In cells K1, K2, and K3, enter 0, 1, and 2, respectively.

3. In cells L1, L2, and L3, enter Iris-setosa, Iris-versicolor, and Iris-virginica. Our worksheet looks like Figure 4-7.

	B	C	D	E	F	G	H	I	J	K	L
1	x1	x2	x3	x4	y					0	Iris-setosa
2	5.1	3.5	1.4	0.2						1	Iris-versicolor
3	4.9	3	1.4	0.2						2	Iris-virginica
4	4.7	3.2	1.3	0.2							
5	4.6	3.1	1.5	0.2							
6	5	3.6	1.4	0.2							
7	5.4	3.9	1.7	0.4							
8	4.6	3.4	1.4	0.3							
9	5	3.4	1.5	0.2							

Figure 4-7. *Iris dataset and table setup*

We would like to make use of the least square method first to obtain the linear function P. Thus, we need to convert the categorical values into numerical values first. Cells K1:L3 define that 0 ➤ Iris-setosa, 1 ➤ Iris-versicolor, and 2 ➤ Iris-virginica.

4. In cell F2, enter the following formula:

```
=IF(A2=L$1,K$1,IF(A2=$L$2,K$2,K$3))
```

5. Autofill cell F2 all the way to F151 (there are 150 data points). Our worksheet should look like Figure 4-8.

▲	A	B	C	D	E	F	G	K	L
1	Iris type	x1	x2	x3	x4	y		0	Iris-setosa
2	Iris-setosa	5.1	3.5	1.4	0.2	0		1	Iris-versicolor
3	Iris-setosa	4.9	3	1.4	0.2	0		2	Iris-virginica
4	Iris-setosa	4.7	3.2	1.3	0.2	0			
5	Iris-setosa	4.6	3.1	1.5	0.2	0			
6	Iris-setosa	5	3.6	1.4	0.2	0			
7	Iris-setosa	5.4	3.9	1.7	0.4	0			
8	Iris-setosa	4.6	3.4	1.4	0.3	0			
9	Iris-setosa	5	3.4	1.5	0.2	0			
10	Iris-setosa	4.4	2.9	1.4	0.2	0			
11	Iris-setosa	4.9	3.1	1.5	0.1	0			
12	Iris-setosa	5.4	3.7	1.5	0.2	0			
13	Iris-setosa	4.8	3.4	1.6	0.2	0			
14	Iris-setosa	4.8	3	1.4	0.1	0			

Figure 4-8. *Iris dataset categorized*

6. In cells K5:K9, enter b, w1, w2, w3, and w4, respectively. These represent the coefficients (weights) and intercept of the linear function P.

7. In cells L5:L9, enter 5, 4, 3, 2, and 1, respectively. We are going to make use of the functions INDEX and LINEST to find the coefficients and intercept for the linear function P. The function LINEST returns an array containing five elements. The order of the elements in the array is w4, w3, w2, w1, and b.

8. In cell M5, enter the formula:

```
=INDEX(LINEST(F$2:F$151,B$2:E$151,TRUE,TRUE),1,L5)
```

9. Autofill cells M5 to M9. Part of our worksheet should look like Figure 4-9. Cells M5:M9 contain the coefficients of the linear function P. Our next step is to apply P to obtain the predicted value for each data point. We will store such predicted values in column G.

K	L	M
0	Iris-setosa	
1	Iris-versicolor	
2	Iris-virginica	
b	5	0.192083995
w1	4	-0.109741463
w2	3	-0.044240447
w3	2	0.227001382
w4	1	0.60989412

Figure 4-9. *Using least square method to obtain function coefficients*

10. Enter "Numerical prediction" in cell G1. In cell G2, enter the following formula:

```
=MMULT(B2:E2, M$6:M$9)+ M$5
```

This formula represents the expression $w_1x_1 + w_2x_2 + \ldots + w_nx_n + b$. Because B2:E2 is a horizontal array while M6:M9 is a vertical array, the multiplication of the two arrays can only be achieved by the function MMULT which performs a matrix-like multiplication. A similar function, SUMPRODUCT, cannot be applied here. This formula can be replaced by a clumsier one (which is not recommended):

```
=M$6*B2+M$7*C2+M$8*D2+M$9*E2+M$5
```

11. Autofill cells G2 to G152. Once we have the predicted values for each data point, what we need next is to find cutoffs to classify each predicted value into a specific iris type.

12. In cells K11, L11, and M11, enter "mean", "sample number", and "cutoff", respectively (no quotation marks).

13. In cell K12, enter the following formula:

```
=AVERAGEIFS(G$2:G$151,F$2:F$151,K1)
```

This formula computes the mean of the predicted values for Iris-setosa.

14. Autofill K12 to K14. The means of the predicted values for Iris-versicolor and Iris-virginica are stored in cells K13 and K14, respectively.

15. In cell L12, enter the formula:

    =COUNTIFS(F$2:F$151,K1)

    This formula counts the number of Iris-setosa in the training dataset.

16. Autofill L12 to L14. Cells L13 and 14 store the number of Iris-versicolor and Iris-virginica, respectively.

17. In cell M12, enter the formula to compute the first cutoff:

    =(K12*L12+K13*L13)/(L12+L13)

18. Autofill M12 to M13. Do not autofill to cell M14. Based on the two cutoffs, each predicted numerical value can be converted into an iris type. Part of our worksheet looks like Figure 4-10.

	G	H	I	J	K	L	M
1	Numerical classification				0	Iris-setosa	
2	-0.082658272				1	Iris-versicolor	
3	-0.038589756				2	Iris-virginica	
4	-0.048189691						
5	0.012608776				b	5	0.192083995
6	-0.076108171				w1	4	-0.109741463
7	0.056802348				w2	3	-0.044240447
8	0.037625916				w3	2	0.227001382
9	-0.044559943				w4	1	0.60989412
10	0.02070502						
11	-0.081303075				mean	sample number	cutoff
12	-0.101728663				-0.027351429	50	0.583253347
13	8.84876E-05				1.193858122	50	1.513675714
14	-0.088605022				1.833493306	50	

Figure 4-10. *Cutoffs computed*

19. In cell H1, type "Type classification".

20. In cell H2, enter the following formula and then autofill from H2 to H152:

    ```
    =IF(G2<M$12,L$1,IF(G2<M$13,L$2,L$3))
    ```

    This formula converts each numerical value in column G into an iris type. A glance of the result would tell that most iris types are classified into the right group.

21. To count how many iris data points are incorrectly classified, enter "Difference" in cell I1.

22. In I2, enter the following formula:

    ```
    =IF(A2=H2,0,1)
    ```

    This formula compares the known iris type with the classified iris type. If a mismatch is found, 1 is returned.

23. Autofill cells I2 to I152.

24. Select cells K16 and L16, merge these two cells, and type inside the text "Difference =" in the merged cell.

25. In cell M16, enter the formula =SUM(I2:I151). Part of our formula should look like Figure 4-11.

H	I	J	K	L	M
Type classification	Difference		0	Iris-setosa	
Iris-setosa	0		1	Iris-versicolor	
Iris-setosa	0		2	Iris-virginica	
Iris-setosa	0				
Iris-setosa	0		b	5	0.192084
Iris-setosa	0		w1	4	-0.10974
Iris-setosa	0		w2	3	-0.04424
Iris-setosa	0		w3	2	0.227001
Iris-setosa	0		w4	1	0.609894
Iris-setosa	0				
Iris-setosa	0		mean	sample number	cutoff
Iris-setosa	0		-0.0274	50	0.58325
Iris-setosa	0		1.19386	50	1.51368
Iris-setosa	0		1.83349	50	
Iris-setosa	0				
Iris-setosa	0			Difference=	4

Figure 4-11. *LDA analysis on the Iris training dataset*

This LDA analysis is quite successful, though there are still four iris data points incorrectly classified. Note that it is common for a data mining method to miss a few data points on the training dataset. Generally speaking, we should not use the same training dataset to assess how good a model can perform, however. A model with a perfect classification or prediction on the same training dataset might suggest an overfit, which indicates that the model is too specific to the training data but is likely to fail on unknown scoring data.

If we assign 0 to Iris-virginica, 1 to Iris-versicolor, and 2 to Iris-setosa in cells K1:L3, we will get a very similar result. However, if we assign 0 to Iris-virginica, 1 to Iris-setosa, and 2 to Iris-versicolor in cells K1:L3, we will find that the classification is totally messed up. This is a disadvantage of carrying out LDA analysis in Excel.

With the first model constructed, that is, the initial linear function P found, we can predict the iris type based on x1, x2, x3, and x4. We can view such predictions in the worksheet iris-LINEST of the file Chapter4-1b.xlsx.

Equation (4-3) requests to maximize the ratio of Sg/Sw. For this purpose, we need to make use of Solver. Note, we can obtain an optimized parameter set by using Solver directly without using LINEST first. However, in LDA, it is a good idea to use least square method first before applying Solver.

Continue our learning through the following step-by-step instructions:

26. In cells K18, K19, and K20, type "Inter-group variance", "Within-group variance", and "inter/within ratio", respectively.

27. Enter the following formula in cell L18:

    ```
    =(K12-AVERAGE(K12:K14))^2+(K13-AVERAGE(K12:K14))^2+(K14-AVERAGE(K12:K14))^2
    ```

    This formula computes the variance among group means, that is, the Sg as shown in Equation (4-1). The larger this value, the higher the separability among groups. Note, cells K12, K13, and K14 reference the within-group means.

28. We must compute the variance of each data point in their belonging group first before computing the within-group variance. Enter "Within-g variances" in cell J1, then enter the following formula in cell J2:

    ```
    =IF(F2=K$1,(G2-K$12)^2,IF(F2=K$2, (G2-K$13)^2,(G2-K$14)^2))
    ```

    In this formula, column F is used to determine which group each data point lies in. Note that column F represents the true group each data point belongs. LDA is a supervised classification method; it is more accurate to use column F than column G here. If F2 = K1, then the within-group variance of the corresponding data point is computed by the expression (G2-K12)^2. If not but F2 = K2, the variance is computed via the expression (G2-K13)^2; otherwise, use the expression (G2-K14)^2.

29. Autofill cells J2 to J152.

30. Enter the formula =SUM(J1:J152) in cell L19. This is the S_w value as shown in Equation (4-2).

31. Enter the formula =L18/L19 in cell L20. This is the Sg(P)/Sw(P) that needs to be maximized by Solver. By now, part of our worksheet looks like Figure 4-12.

	J	K	L	M
1	Within-g variances		0 Iris-setosa	
2	0.003058847		1 Iris-versicolor	
3	0.0001263		2 Iris-virginica	
4	0.000434233			
5	0.001596818	b	5	0.192084
6	0.00237722	w1	4	-0.10974
7	0.007081858	w2	3	-0.04424
8	0.004222055	w3	2	0.227001
9	0.000296133	w4	1	0.609894
10	0.002309422			
11	0.00291078	mean	sample number	cutoff
12	0.005531973	-0.027351429	50	0.58325
13	0.000752949	1.193858122	50	1.51368
14	0.003752003	1.833493306	50	
15	0.005547758			
16	0.039858672	Difference=		4
17	0.000265337			
18	4.41796E-05	Inter-group variance	1.787743021	
19	3.22916E-05	Within-group variance	3.65508569	
20	2.84519E-05	inter/within ratio	0.489111111	

Figure 4-12. *Variances computed*

32. Copy the content of cells M5:M9 to cells N5:N9 by value only. Remember to make use of the Excel feature "Paste Special". We are

going to use Solver to optimize these values in cells M5:M9. Before doing that, we would like to make a copy of them so that we can tell the changes after using Solver. At this point, take a look at the values in cells L18 and L19; try to remember them.

33. Click the menu tab Data; locate the Solver as shown in Figure 4-13.

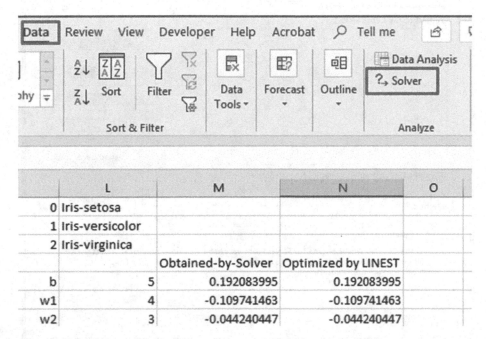

Figure 4-13. *Using Excel Solver*

34. Click Solver. On the upcoming menu, specify the values exactly as shown in Figure 4-14. Set Objective indicates the value to be optimized. In this case, we need the value in L20 to be maximized. The Solver is requested to maximize L20 by changing values in cells M5:M9. The selected Solving Method is GRG Nonlinear.

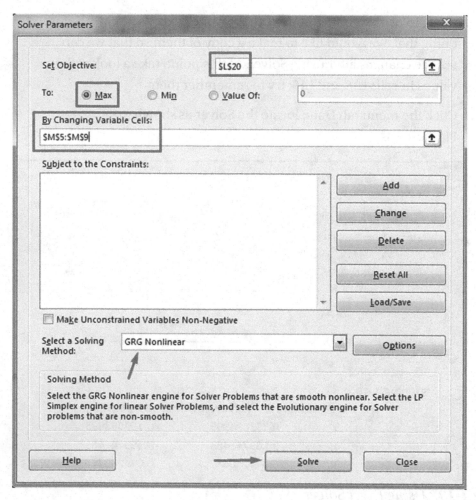

Figure 4-14. *Using Excel Solver to optimize the parameters for function P*

35.   After clicking Solve, click OK on the upcoming menu as
demonstrated in Figure 4-15.

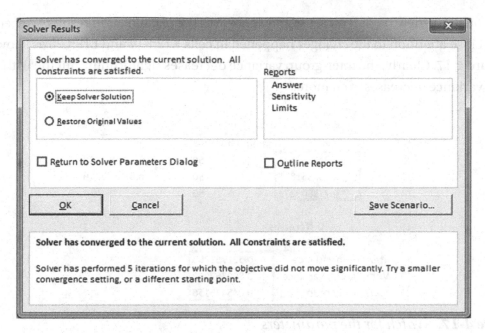

Figure 4-15. *Keep Solver Solution*

The values inside cells M5:M9 are changed, and the value in cell M16 is changed to 2 from 4. This is shown in Figure 4-16.

K	L	M	N
0	Iris-setosa		
1	Iris-versicolor		
2	Iris-virginica		
		Obtained-by-Solver	Optimized by LINEST
b	5	0.192083994	0.192083995
w1	4	-0.060753681	-0.109741463
w2	3	-0.114783909	-0.044240447
w3	2	0.162026358	0.227001382
w4	1	0.211629585	0.60989412
mean	sample number	cutoff	
-0.215536129	50	0.134407849	
0.484351827	50	0.631577549	
0.778803272	50		
Difference =		2	

Figure 4-16. *Solver further improves the LDA model*

Though the new parameter set slightly improves the classification results, we need to pay close attention to the changes happened in cells K12:K14 and L18:L20 as shown in Figure 4-17. Clearly, the inter-group variance decreases significantly, but the within-group variance decreases even more.

	K	L	M
11	mean	sample number	cutoff
12	-0.215536129	50	0.134407849
13	0.484351827	50	0.631577549
14	0.778803272	50	
15			
16	Difference =		2
17			
18	Inter-group variance	0.521751883	
19	Within-group variance	0.808367262	
20	inter/within ratio	0.645439156	

Figure 4-17. *Watch for the parameters*

To assess how good our model is, we need to conduct cross-validation on our model and learn receiver operating characteristic (ROC) curve analysis. This will be the topic of the next chapter.

Reinforcement Exercises

Please open the file Chapter4-HW.xlsx and try the LDA exercise inside. Should you experience any difficulty, please refer to the Excel file named Chapter4-HW-withAnswers.xlsx.

Review Points

1. General understanding of LDA

2. Solver, including its installation and use

3. Excel functions INDEX and LINEST

4. Excel functions IF, MMULT, COUNTIFS, and AVERAGEIFS

CHAPTER 5

Cross-Validation and ROC

Please download the sample Excel files from https://github.com/hhohho/Learn-Data-Mining-through-Excel-2 for this chapter's exercises.

General Understanding of Cross-Validation

A prediction model should be validated before it can be successfully applied to scoring data. Using the same training dataset as the testing dataset to assess the constructed model is not a good way to validate the model. Such a validating strategy is called residual analysis. It compares the difference (so-called residual) between the actual output and the predicted output. In Chapter 4, we applied residual analysis to assess the quality of our constructed LDA model.

Residual analysis cannot tell how well the model can perform on the scoring data, the future data with unknown target values. One solution is to leave part of the original training dataset unused as the testing subset. In many occasions, the testing subset is also called validation subset, though they can be further distinguished. In this book, let's treat them the same. As the model is not trained with the testing subset, such a testing subset can act as a scoring dataset and can also tell us how well the model can perform with unknown data. Such a validation strategy is called cross-validation.

There are several cross-validation strategies. The simplest one is the holdout method in which the original training dataset is divided into two subsets: training and testing. The model is trained on the training subset but assessed on the testing subset. This method is more reliable than residual analysis, but its performance heavily relies on what is inside the training subset and what is inside the testing subset, that is, how to divide the original training dataset can significantly impact the model performance. A consensus in machine learning community is that the training subset should be larger

© Hong Zhou 2023
H. Zhou, *Learn Data Mining Through Excel*, https://doi.org/10.1007/978-1-4842-9771-1_5

than the testing subset, usually at least twice as large. In addition, the training subset should be as large as possible as long as there are enough data points inside the testing subset to evaluate the model.

K-fold cross-validation is an improved version over the holdout method. In k-fold cross-validation, the original training data is randomly divided into k subsets, and the holdout assessment is performed k times. In each holdout assessment, one subset is used as the testing data for validation; all other k-1 subsets are grouped together for training. In k-fold cross-validation, each data point is used once in the testing subset but k-1 times in the training subset. The model's performance is based on the overall quality of the k assessments.

One widely used cross-validation method is the leave-one-out (LOO) method, which is an extreme version of the k-fold cross-validation. In LOO, the testing data contains only one data point every time; the rest of the data points are grouped together as the training subset.

Learn Cross-Validation Through Excel

When the dataset is large, it is impractical to conduct LOO cross-validation in Excel if no VBA programming is involved. In fact, it is very tedious to carry out a k-fold cross-validation in Excel, too. Therefore, I will only demonstrate how to conduct a holdout cross-validation using Excel. However, file Chapter5-cv-1c.xlsx presents a demonstration of a fivefold cross-validation using LDA.

In the last chapter, we have learned how to carry out linear discriminant analysis on the Iris dataset. In this chapter, the Iris dataset has been divided into two subsets: training and testing. A random algorithm is used to pick about two-third data points as the training dataset, while the rest is the testing dataset.

Open the file Chapter5-cv-1a.xlsx. There are two worksheets: training_dataset, which contains 103 data records, and testing_dataset, which contains 47 data records. Follow what we have learned about LDA in the last chapter; let's make use of the function LINEST to find the parameter set $(w_1, w_2, ...w_n, b)$ for function P in the worksheet training_dataset. Our worksheet should look like Figure 5-1. Note that there are only 103 data points in the training dataset.

	A	B	C	D	E	F	G	H	I	J	K	L	M
1	Iris type	x1	x2	x3	x4	y	Numerical cl	Type classific	Difference		0	Iris-setosa	
2	Iris-setosa	4.6	3.1	1.5	0.2	0	0.03	Iris-setosa	0		1	Iris-versicolor	
3	Iris-setosa	4.6	3.4	1.4	0.3	0	0.06	Iris-setosa	0		2	Iris-virginica	
4	Iris-setosa	4.8	3.4	1.6	0.2	0	0.02	Iris-setosa	0				Obtained-by-LINEST
5	Iris-setosa	5.7	4.4	1.5	0.4	0	-0.05	Iris-setosa	0		b	5	0.35469596
6	Iris-setosa	4.8	3.4	1.9	0.2	0	0.10	Iris-setosa	0		w1	4	-0.178346893
7	Iris-setosa	5.2	4.1	1.5	0.1	0	-0.14	Iris-setosa	0		w2	3	-0.00549763
8	Iris-setosa	5.1	3.8	1.5	0.3	0	0.00	Iris-setosa	0		w3	2	0.267646734
9	Iris-setosa	5.1	3.3	1.7	0.5	0	0.17	Iris-setosa	0		w4	1	0.579921945
10	Iris-setosa	5.4	3.4	1.5	0.4	0	0.01	Iris-setosa	0				
11	Iris-setosa	4.9	3.1	1.5	0.1	0	-0.08	Iris-setosa	0		mean	ample numbe	cutoff
12	Iris-setosa	4.6	3.2	1.4	0.2	0	0.01	Iris-setosa	0		-0.02	35	0.557350986
13	Iris-setosa	5.4	3.9	1.7	0.4	0	0.06	Iris-setosa	0		1.17	33	1.526466881
14	Iris-setosa	4.8	3	1.4	0.1	0	-0.09	Iris-setosa	0		1.86	35	
15	Iris-setosa	5.1	3.5	1.4	0.3	0	-0.03	Iris-setosa	0				
16	Iris-setosa	5.7	3.8	1.7	0.3	0	-0.05	Iris-setosa	0			difference=	2

Figure 5-1. *Apply LDA on the training dataset*

Again, the LDA works well on the Iris dataset. There are only two false predictions. However, the validation results in Figure 5-1 were obtained through a residual analysis.

Follow these instructions to complete the cross-validation procedure in Excel:

1. In the worksheet training_dataset, the parameter set (w_1, w_2, ...w_n, b) is stored in cells M5:M9, while the cutoffs are referenced by cells M12 and M13. Select cells K1:M16 and copy them.

2. Click the testing_dataset worksheet to open it.

3. In the worksheet testing_dataset, right-click on cell K1 ➤ apply the Paste Special skill to paste values into cell K1. Remember to paste values only.

    Our testing_dataset worksheet should look like Figure 5-2.

	A	B	C	D	E	F	G	H	I	J	K	L	M
1	Iris type	x1	x2	x3	x4						0	Iris-setosa	
2	Iris-setosa	5	3	1.6	0.2						1	Iris-versicolor	
3	Iris-setosa	5.2	3.5	1.5	0.2						2	Iris-virginica	
4	Iris-setosa	4.9	3.1	1.5	0.1								Obtained-by-LINEST
5	Iris-setosa	5	3.5	1.3	0.3						b	5	0.35469596
6	Iris-setosa	5.1	3.8	1.9	0.4						w1	4	-0.178346893
7	Iris-setosa	5	3.3	1.4	0.2						w2	3	-0.00549763
8	Iris-setosa	5	3.4	1.5	0.2						w3	2	0.267646734
9	Iris-setosa	4.3	3	1.1	0.1						w4	1	0.579921945
10	Iris-setosa	5.8	4	1.2	0.2								
11	Iris-setosa	5.1	3.7	1.5	0.4						mean	sample numl	cutoff
12	Iris-setosa	5.1	3.4	1.5	0.2						-0.02	35	0.557350986
13	Iris-setosa	5	3.5	1.6	0.6						1.173	33	1.526466881
14	Iris-setosa	4.9	3.1	1.5	0.1						1.86	35	
15	Iris-setosa	4.5	2.3	1.3	0.3								
16	Iris-setosa	5.4	3.7	1.5	0.2							Difference=	2

Figure 5-2. *Copy the LDA parameter set to the testing worksheet*

4. In cells G1:I1, type "Numerical classification", "Type classification", and "Difference", respectively.

5. In cell G2, enter the following formula:

    ```
    =MMULT(B2:E2, M$6:M$9)+ M$5
    ```

6. Autofill from G2 to G48.

7. In cell H2, enter the following formula:

    ```
    =IF(G2<M$12,L$1,IF(G2<M$13,L$2,L$3))
    ```

8. Autofill from H2 to H48.

9. In cell I2, enter the formula =IF(A2=H2,0,1), hit Enter, and then autofill from I2 to I48.

10. At last, in cell M16, enter the formula =SUM(I2:I48).

    By now, our testing_dataset worksheet should look like Figure 5-3. The preceding procedure applies the parameter set (including coefficients, intercept, and cutoffs) obtained through the training dataset to the testing dataset. It compares the actual Iris types (in column A) with the predicted types (in column H). There are two samples falsely predicted. As there are no common data points between the training and testing datasets, this validation through the testing dataset is a bona fide cross-validation.

	A	B	C	D	E	F	G	H	I	J	K	L	M
1	Iris type	x1	x2	x3	x4		Numerical classifie	Type classifica	Difference		0	Iris-setosa	
2	Iris-setosa	5	3	1.6	0.2		-0.009312232	Iris-setosa	0		1	Iris-versicolor	
3	Iris-setosa	5.2	3.5	1.5	0.2		-0.074495099	Iris-setosa	0		2	Iris-virginica	
4	Iris-setosa	4.9	3.1	1.5	0.1		-0.076784174	Iris-setosa	0				Obtained-by-LINEST
5	Iris-setosa	5	3.5	1.3	0.3		-0.034362872	Iris-setosa	0		b	5	0.35469596
6	Iris-setosa	5.1	3.8	1.9	0.4		0.164733384	Iris-setosa	0		w1	4	-0.178346893
7	Iris-setosa	5	3.3	1.4	0.2		-0.064490868	Iris-setosa	0		w2	3	-0.00549763
8	Iris-setosa	5	3.4	1.5	0.2		-0.038275957	Iris-setosa	0		w3	2	0.267646734
9	Iris-setosa	4.3	3	1.1	0.1		-0.076284968	Iris-setosa	0		w4	1	0.579921945
10	Iris-setosa	5.8	4	1.2	0.2		-0.26454607	Iris-setosa	0				
11	Iris-setosa	5.1	3.7	1.5	0.4		0.058224454	Iris-setosa	0		mean	sample num	cutoff
12	Iris-setosa	5.1	3.4	1.5	0.2		-0.056110647	Iris-setosa	0		-0.02	35	0.557350986
13	Iris-setosa	5	3.5	1.6	0.6		0.219907731	Iris-setosa	0		1.173	33	1.526466881
14	Iris-setosa	4.9	3.1	1.5	0.1		-0.076784174	Iris-setosa	0		1.86	35	
15	Iris-setosa	4.5	2.3	1.3	0.3		0.06140773	Iris-setosa	0				
16	Iris-setosa	5.4	3.7	1.5	0.2		-0.111264003	Iris-setosa	0			Difference=	2

Figure 5-3. *Cross-validation with the testing dataset*

11. We can certainly further optimize the LDA model in the training_ dataset worksheet. Follow the procedure we have learned in the last chapter to apply Solver to the training dataset.

Recall in the last chapter, after using Solver to optimize the model, the "Inter-group variance" significantly decreased. However, working with this smaller training dataset, after applying Solver to optimize the model, the "Inter-group variance" significantly increased. This is a good example illustrating that data sample does matter in data mining. Part of the result is shown in Figure 5-4.

K	L	M	N
0	Iris-setosa		
1	Iris-versicolor		
2	Iris-virginica		
		Obtained-by-Solver	Optimized by LINEST
b	5	0.35469596	0.35469596
w1	4	-0.175593016	-0.178346893
w2	3	-0.260502399	-0.00549763
w3	2	0.409467301	0.267646734
w4	1	0.529952989	0.579921945
mean	sample number	cutoff	
-0.692021015	35	0.137766368	
1.017843895	33	1.424748555	
1.80840152	35		
	Difference=	2	
Inter-group variance	3.266910741		
Within-group variance	3.366757214		
inter/within ratio	0.970343429		

Figure 5-4. *Parameter set obtained through Solver in the training dataset*

12. Apply the parameter set to the testing subset (copying values in cells M5:M9 which are optimized by Solver in the training_dataset worksheet to the testing_dataset worksheet).

13. In L16, enter "Cross validation missed =". In L17, enter "Cross validation accuracy=", and in M17, enter the formula =1-M16/47. Our result looks like Figure 5-5.

The complete results are available in the file Chapter5-cv-1b.xlsx.

H	I	J	K	L	M
Type classification	Difference		0	Iris-setosa	
Iris-setosa	0		1	Iris-versicolor	
Iris-setosa	0		2	Iris-virginica	
Iris-setosa	0				Optimized by Solver
Iris-setosa	0		b	5	0.35469596
Iris-setosa	0		w1	4	-0.175593016
Iris-setosa	0		w2	3	-0.260502399
Iris-setosa	0		w3	2	0.409467301
Iris-setosa	0		w4	1	0.529952989
Iris-setosa	0				
Iris-setosa	0		mean	sample number	cutoff
Iris-setosa	0		-0.02285	35	0.557350986
Iris-setosa	0		1.172716	33	1.526466881
Iris-setosa	0		1.860004	35	
Iris-setosa	0				
Iris-setosa	0			Cross validation missed =	1
Iris-versicolor	0			Cross validation accuracy=	97.87%
Iris-versicolor	0				

***Figure 5-5.** Cross-validation with testing dataset*

Bear in mind that in cross-validation, only part of the original training dataset is used to construct the model, while the rest is used to test the model performance. Once the data mining method, for example, LDA, is confirmed to have acceptable performance, it is better for the final model (or the parameter set) to be generated based on the whole original training dataset using LDA method.

General Understanding of ROC Analysis

Receiver operating characteristic (ROC) curve analysis is another commonly used method to assess the performance of a model. Different from cross-validation, ROC analysis works only for the case when the target variable has exactly two distinct values, such as true and false, or positive and negative. For this reason, ROC analysis cannot be used to assess the preceding Iris dataset LDA model.

ROC analysis is widely used in biomedical model assessments, where a diagnostic test must determine if a sample is positive or negative regarding a disease. The following scenario helps explain what ROC analysis can be applied.

Suppose our research has discovered that certain blood characteristics are well correlated with the identification of a certain type of disease. Let's name these characteristics C1, C2, C3, and C4 and the disease D. We have developed an LDA model based on C1, C2, C3, and C4 to determine the presence of the disease in patients. A perfect model would clearly separate the group with the disease from the group without the disease in a large population. Unfortunately, clear separation doesn't exist in reality because of classification errors. In the group having the disease, some are correctly classified as positive (true positive, or TP), while some are incorrectly classified as negative (false negative, or FN). Similarly, in the group without the disease, some are correctly classified as negative (true negative, or TN), while some are mistakenly classified as positive (false positive, or FP). By adjusting the cutoff or other related assessment metric, the numbers of TP, FN, FP, and TN can vary.

The following concepts are critical in understanding and applying ROC analysis:

- Sensitivity is the probability that a test result will be positive when the disease is present. It is also called true positive rate and is defined as TP / (TP + FN). It only measures the percentage of diseased individuals that can be correctly classified by the test.

- Specificity is the probability that a test result will be negative when the disease is not present. It is also called true negative rate and is computed as TN / (TN + FP). It only measures the percentage of un-diseased individuals that can be correctly classified by the test.

- False positive rate is the probability that a test result will be positive when the disease is not present. False positive rate = 1-specificity.

- Positive predictive value is the probability that the disease is present when the test is positive, expressed as TP / (TP + FP).

- Negative predictive value is the probability that the disease is absent when the test is negative, expressed as TN / (TN + FN).

Consider the extreme situation when the test determines every individual as positive. In this case, FN = 0 and TN = 0, specificity = 0 but sensitivity = 100%.

In another extreme situation, when the test determines every individual as negative, specificity = 100%, while sensitivity = 0.

By adjusting the cutoff, both sensitivity and specificity can vary. In most cases, we will observe when sensitivity goes up, specificity goes down. A good model has a cutoff point that can keep both sensitivity and specificity high. However, some tests may emphasize on high positive predictive values or high negative predictive values.

Learn ROC Analysis Through Excel

Data inside the file Chapter5-roc-1a.xlsx are simulated. In the file, TP, FN, FP, and TN change along with different cutoff values, as shown in Figure 5-6. Though there are 19 records in the file, Figure 5-6 only displays 11 records. Records 8–15 are hidden.

	A	B	C	D	E
1	True Positive	Fasle Negative	False Positive	True Negative	
2	TP	FN	FP	TN	Cut-off
3	552	48	384	16	1
4	552	48	382	18	1.5
5	552	48	380	20	2
6	547	53	360	40	2.5
7	546	54	352	48	3
16	390	210	48	352	7.5
17	342	258	28	372	8
18	312	288	24	376	8.5
19	258	342	20	380	9
20	243	357	18	382	9.5
21	183	417	10	390	10

Figure 5-6. *Chapter5-roc-1a.xlsx training dataset*

We want to draw a chart to view how both sensitivity and specificity change along with different cutoff values. Follow these instructions:

1. Enter "sensitivity" and "specificity" in cells F2 and G2, respectively.

2. In cell F3, enter the formula =A3/(A3+B3). This formula computes sensitivity when cutoff = 1. Autofill from F3 to F21.

3. In cell G3, enter the formula =D3/(C3+D3). This formula calculates specificity when cutoff = 1. Autofill from G3 to G21.

4. Select cells E2:G21; click the menu tab Insert ➤ Chart ➤ select Scatter with Smooth Lines and Markers. A scatter chart shows up in our worksheet.

5. We can add axis titles and chart title as desired. The vertical axis's bound maximum is 1.2 by default. I prefer setting it to 1.0, so I double-click on the value 1.2 and then change it to 1.0 when the menu for Axis Options shows up.

Our chart may look similar to Figure 5-7 (related data included). Note that the two curves cross each other at a point. The intersection point marks a cutoff value at which the sensitivity equals the specificity.

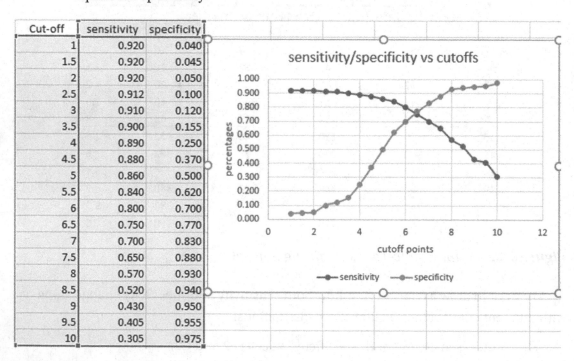

Cut-off	sensitivity	specificity
1	0.920	0.040
1.5	0.920	0.045
2	0.920	0.050
2.5	0.912	0.100
3	0.910	0.120
3.5	0.900	0.155
4	0.890	0.250
4.5	0.880	0.370
5	0.860	0.500
5.5	0.840	0.620
6	0.800	0.700
6.5	0.750	0.770
7	0.700	0.830
7.5	0.650	0.880
8	0.570	0.930
8.5	0.520	0.940
9	0.430	0.950
9.5	0.405	0.955
10	0.305	0.975

Figure 5-7. *Sensitivity/specificity vs. cutoffs*

The next step is to draw a ROC curve. ROC curve is graphed as false positive rate vs. true positive rate, that is, 1-specificity vs. sensitivity. To make it more viewable, the values are all multiplied by 100. So, the ROC curve is in fact (100-100*specificity) vs. (100*sensitivity).

In a ROC curve chart, the x-axis is 100-specificity. Therefore, the specificity values should be arranged in descending order. As the current specificity values are in ascending order, we need to flip-flop the specificity values. Accordingly, we need to flip-flop the sensitivity values, too. Continue following the instructions below:

6. Select cells E2:G21; use Paste Special to paste values only into cell A25.

7. Select cells A25:C44; click menu tab Data ➤ Sort.

8. On the upcoming menu, sort the data by Cut-off and the order is Largest to Smallest, as shown in Figure 5-8. Make sure to check "My data has headers". Click OK.

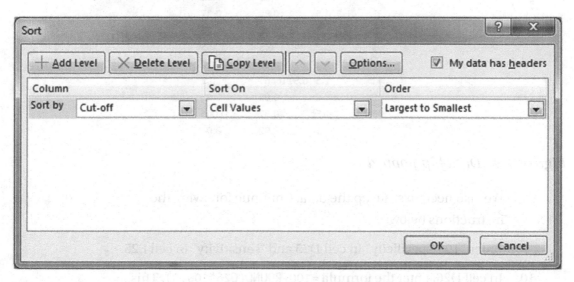

Figure 5-8. Sort data by Cut-off

Part of our worksheet looks like Figure 5-9.

Excel can sort the selected cells only while leaving all other cells intact. This is a great feature of Excel. If the to-be sorted cells have formula referencing other cells outside the sorting area, sorting can mess up some calculations. This is why we need to copy E2:G21 by values first and then sort the copied values.

	A	B	C
25	Cut-off	sensitivity	specificity
26	10	0.305	0.975
27	9.5	0.405	0.955
28	9	0.43	0.95
29	8.5	0.52	0.94
30	8	0.57	0.93
31	7.5	0.65	0.88
32	7	0.7	0.83
33	6.5	0.75	0.77
34	6	0.8	0.7
35	5.5	0.84	0.62
36	5	0.86	0.5
37	4.5	0.88	0.37
38	4	0.89	0.25
39	3.5	0.9	0.155
40	3	0.91	0.12
41	2.5	0.9116667	0.1
42	2	0.92	0.05
43	1.5	0.92	0.045
44	1	0.92	0.04

Figure 5-9. *Data flip-flopped*

We also need to scale up the data. Continue following the instructions below:

9. Enter "100-specificity" in cell D25 and "sensitivity" in cell E25.

10. In cell D26, enter the formula =100-ROUND(C26*100,0). This formula multiplies the existing specificity value by 100, rounds it, and subtracts 100 by it.

11. Autofill cells D26 to D44.

12. In cell E26, enter the formula =ROUND(100*B26,0).

13.   Autofill cells E26 to E44.

14.   Right-click on the row number 26 and select "Insert". This action inserts a blank row. The original cell D26 becomes D27 now.

15.   Enter 0 in both cells D26 and E26.

Part of our worksheet should look like Figure 5-10.

	A	B	C	D	E
25	Cut-off	sensitivity	specificity	100-specificity	sensitivity
26				0	0
27	10	0.305	0.975	2	31
28	9.5	0.405	0.955	4	41
29	9	0.430	0.950	5	43
30	8.5	0.520	0.940	6	52
31	8	0.570	0.930	7	57
32	7.5	0.650	0.880	12	65
33	7	0.700	0.830	17	70
34	6.5	0.750	0.770	23	75
35	6	0.800	0.700	30	80
36	5.5	0.840	0.620	38	84
37	5	0.860	0.500	50	86

Figure 5-10. *Set up data for ROC curve*

Now comes the last step to draw the ROC curve:

16.   Select cells D25:E45.

17.   Click menu tab Insert ➤ Chart ➤ Scatter with Smooth Lines.

18.   Add chart title and axis titles as desired. Our ROC chart may look similar to Figure 5-11.

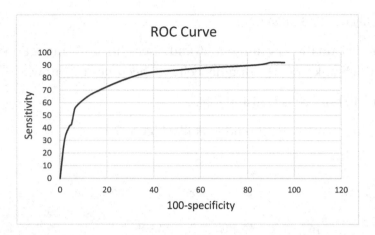

Figure 5-11. *The ROC curve*

The size of the area under the ROC curve measures the quality of a model. A large value of the area indicates that the model can have both good sensitivity and specificity. In a perfect model, both the sensitivity and specificity are 100%. This means, when 100-specificity equals 0, the sensitivity is 100. In another word, a perfect model has a ROC curve that goes through the upper-left point. This rarely happens, however.

The file Chapter5-roc-1b.xlsx presents the results of the preceding exercise.

This wraps up another chapter. Why do we need to learn cross-validation and ROC before we learn other data mining methods? Well, once we have the knowledge about cross-validation and ROC, we can practice them when we learn other data mining methods. After this chapter, we will learn logistic regression, k-nearest neighbors, Naïve Bayes, decision trees, etc. At one point, we may want to assess how good our models are, and that is when we can practice cross-validation and/or ROC curve analysis.

Reinforcement Exercises

This chapter has two reinforcement exercises, one for cross-validation and one for ROC curve analysis. It is a good idea to go through both to reinforce your understanding. You should find four Excel files for the two reinforcement exercises:

1. Chapter5-HW-CrossValidation.xlsx

2. Chapter5-HW-CrossValidation-Answers.xlsx

3. Chapter5-roc-HW.xlsx

4. Chapter5-roc-HW-withAnswers.xlsx

Review Points

1. Residual analysis

2. Testing dataset

3. The holdout method of cross-validation

4. K-fold cross-validation and LOO cross-validation

5. What is ROC

6. True positive, false positive, true negative, and false negative

7. Sensitivity and specificity

8. ROC curve

9. Excel skills and functions:

    a. Sort

    b. Scatter chart

    c. Functions IF, ROUND, and MMULT

CHAPTER 6

Logistic Regression

Please download the sample Excel files from `https://github.com/hhohho/Learn-Data-Mining-through-Excel-2` for this chapter's exercises.

General Understanding

Logistic regression can be thought of as a special case of linear regression when the predicted outcomes are categorical. If there are only two outcomes, the logistic regression is called binomial logistic regression which is the most popular one. If there are more than two outcomes, the logistic regression is called multinomial logistic regression. In this chapter, we are learning binomial logistic regression in Excel.

Binomial logistic regression can also be thought of as a special case of LDA, but its mechanism of achieving good separability between data groups is different from that of LDA.

The name of logistic regression comes from the fact that logistic function is used in the model. Equation (6-1) shows the logistic function in its general form (where a, b, and c are all positive numbers).

$$f(x) = \frac{c}{1 + a \cdot b^{-kx}} \tag{6-1}$$

The well-known sigmoid function is a special case of the logistic function.

$$s(x) = \frac{1}{1 + e^{-x}} \tag{6-2}$$

© Hong Zhou 2023
H. Zhou, *Learn Data Mining Through Excel*, https://doi.org/10.1007/978-1-4842-9771-1_6

Logistic regression is a statistical model, capable of estimating the probability of the occurrence of an event. Let 1 represent the case when the event happens and 0 represent the case when the event does not happen. Thus, P(1) is the probability of the event and P(0) = 1- P(1). Note, the probability must be between 0 and 1 inclusively. Finally, P(1)/(1-P(1)) is the odds for the event to happen.

Suppose the occurrence of the event is dependent on the variables $x_1, x_2, \cdots, x_n$. In logistic regression, it is believed that the log of the odds is a linear function of x_i, as depicted in Equation (6-3).

$$\ln\left(\frac{P(1)}{1-P(1)}\right) = m_1 x_1 + m_2 x_2 + \cdots + m_n x_n + b \tag{6-3}$$

Solving for P(1), we can get the logistic function as shown in Equation (6-4).

$$P(1) = \frac{1}{1 + e^{-(m_1 x_1 + m_2 x_2 + \cdots + m_n x_n + b)}} \tag{6-4}$$

Logistic regression does not maximize the probability of P(1) by optimizing the coefficients $m_1, m_2, \cdots, m_n, b$ (here, let's take b as a coefficient; later we will know that b is also called "bias" in machine learning community). Instead, it tries to maximize the likelihood associated with each data sample. What then is likelihood? Simply speaking, likelihood is the probability of a sample to match the actual event outcome (either 1 or 0). Still confused? Let me use one example to explain what is likelihood.

Assume there are two data records A and B in the training dataset. A has the event happened, that is, its event outcome is 1, and B does not have the event happened, that is, its event outcome is 0. Through the logistic function, the P(1) for A and the P(1) for B are calculated as 0.8 and 0.6, respectively. Because A's actual event outcome is 1, A's likelihood is then its probability for the event to happen, that is, P(1) which is 0.8. On the other hand, because B's actual outcome is 0, its likelihood is its probability for the event not to happen, that is, its matching event outcome becomes 0. Thus, B's likelihood is P(0) instead of P(1), and P(0) = 1 - P(1) = 0.4.

Logistic regression tries to maximize the likelihoods of all samples as a whole. By maximizing the likelihoods, logistic regression achieves good separability between two groups. In implementation, it is common to use the logarithm of a likelihood, either natural logarithm or base 10 logarithm. The advantage of logarithm lies in the fact that it decreases rapidly when the predicted likelihood diverges from the actual event outcome, regardless if the actual event outcome is 1 or 0. Because when the likelihood is 0, the calculation of log(0) will result in an error. Thus, log(0) is treated as a predefined minimum number.

In implementation, logistic regression can also be achieved by minimizing the log loss of likelihoods. The log loss is the negation of the log of a likelihood. Thus, when the actual outcome is 1, the log loss is computed as -log(P(1)), and when the actual outcome is 0, the log loss is computed as -log(1-P(1)). Here, log(0) is treated as a predefined maximum number. Note, the log loss increases rapidly when the predicted likelihood diverges from the actual event outcome. Figure 6-1 shows the log loss when the event outcome is 1. Here, the log(0) is predefined as 0.000001.

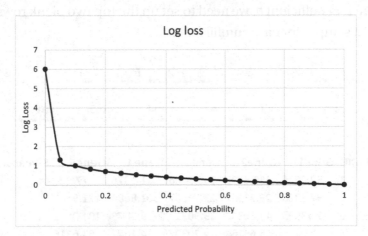

Figure 6-1. *The calculation of log loss when event = 1*

With this understanding, our approach of logistic regression is to use linear regression to obtain a set of coefficients first, then apply Solver to optimize these coefficients by either maximizing the likelihoods as a whole or minimizing the log loss as a whole.

Learn Logistic Regression Through Excel

Data inside the two files Chapter6-1a.xlsx and Chapter6-2a.xlsx are the same. They simulate the following scenario:

> *Suppose there are five genes whose expression levels can be used to predict the five-year survival probability of patients with a certain cancer disease. Value 1 indicates the patient did survive five years, and 0 means they did not.*

As stated earlier, in our logistic regression practices, we are going to apply Solver to optimize the model parameters (coefficients) by either maximizing the likelihoods as a whole or minimizing the log loss as a whole.

By Means of Maximizing Log Likelihoods

Please open the file Chapter6-1a.xlsx in which there are 87 samples and there are 4 blank rows on the top of the worksheet. These blank rows are left there on purpose.

Excel functions INDEX and LINEST will be used again to obtain the coefficients based on the least square method. As the first value returned by LINEST is for Gene 5, and the last value for coefficient b, we need to set up the top two blank rows as shown in Figure 6-2. Such a setup is for easy autofill later.

⁤	A	B	C	D	E	F	G
1		5	4	3	2	1	6
2		m1	m2	m3	m4	m5	b
3							
4							
5	PatientID	Gene1	Gene2	Gene3	Gene4	Gene5	5-year survival
6	1	92.0826	443.3735	350.9466	11.1876	77.926	0
7	2	97.1228	29.21562	2.579007	301.8684	171.9968	0
8	3	7.73995	42.36842	39.90712	9.2879	104.2632	1
9	4	60.2125	25.63164	2.420307	14.1677	94.6316	1
10	5	385.4766	20.32964	5.831751	35.4298	71.0588	0
11	6	476.644	12.48745	4.406125	50.444	104.7838	0

Figure 6-2. *Set up the data for least square method coefficients*

Follow these instructions to exercise logistic regression in Excel:

1. Enter the following formula in both cells B3 and B4:

    =INDEX(LINEST($G$6:$G$92,$B$6:$F$92),1,B1)

    The reason to enter the same formula in two cells is that we want to keep one set of coefficients unchanged while let another set be modified by Solver.

Note, the "known_ys" parameter inside the LINEST function is cells $G$6:$G$92, which have only two values: 1 and 0. This indicates that if our original target values are not 1 and 0, they must be categorized into such.

2. Autofill B3 to G3, and B4 to G4.

3. Enter the text "m1x1+m2x2+...+b" in cell H5. Column H represents the right-side expression of Equation (6-3).

4. In cell H6, enter the following formula and autofill to cell H92:

=SUMPRODUCT($B$4:$F$4,B6:F6)+$G$4

The function SUMPRODUCT computes the product between two arrays. Note that absolute references are used for cell references B4, F4, and G4.

5. Enter the text "P(1)" in cell I5. Column I represents the five-year survival probability computed based on the values in column H.

6. In cell I6, enter the following formula and autofill to cell I92:

=1/(1+EXP(0-H6))

This formula implements Equation (6-4). (0-H6) represents $-(m_1x_1 + m_2x_2 + \cdots + m_nx_n + b)$, while EXP(0-H6) stands for $e^{-(m_1x_1+m_2x_2+\cdots+m_nx_n+b)}$. Function EXP returns the power of the constant e (the base of the natural logarithm). By now, part of our worksheet should look like Figure 6-3.

◢	A	B	C	D	E	F	G	H	I
1		5	4	3	2	1	6		
2		m1	m2	m3	m4	m5	b		
3		-0.00032	0.00301	-0.00513	-0.00051	-0.00074	0.535574025		
4		-0.00032	0.00301	-0.00513	-0.00051	-0.00074	0.535574025		
5	patientID	gene1	gene2	gene3	gene4	gene5	5-year survival	m1x1+m2x2+	P(1)
6	1	92.0826	443.3735	350.9466	11.1876	77.926	0	-0.02392719	0.494018
7	2	97.1228	29.21562	2.579007	301.8684	171.9968	0	0.298006125	0.573955
8	3	7.73995	42.36842	39.90712	9.2879	104.2632	1	0.373754353	0.592366
9	4	60.2125	25.63164	2.420307	14.1677	94.6316	1	0.503845363	0.623363
10	5	385.4766	20.32964	5.831751	35.4298	71.0588	0	0.374075115	0.592443
11	6	476.644	12.48745	4.406125	50.444	104.7838	0	0.296268835	0.57353
12	7	25.89	18.49515	28.72168	6.4724	78.3981	0	0.374183693	0.59247

Figure 6-3. P(1) computed

At this point, it would be interesting to know how well the coefficients generated by the least square method can differentiate the two groups of patients: survived or did not survive five years.

7. Enter "Outcome" in cell J5 and the formula =IF(I6>0.5,1,0) in cell J6. This formula specifies if the probability in column I is greater than 0.5 (here, 0.5 is the cutoff), the predicted outcome of the patient is 1, otherwise 0.

8. Autofill from cell J6 to cell J92.

9. Enter "Difference" in cell K5 and enter the formula =IF(G6=J6,0,1) in cell K6. This formula asserts that if the predicted outcome in column J does match the actual outcome in column G, 0 is returned; otherwise, 1 is returned.

10. Autofill from cell K6 to cell K92.

11. Enter "total diff=" in cell J2, and enter the formula =SUM(K6:K92) in cell K2. By counting how many 1s there are in the array K6:K92, we can tell how many predictions do not match the actual outcomes.

Note that the value in cell K2 is 53. This indicates that 53 out of 87 samples are falsely classified. This is illustrated in Figure 6-4, which displays part of our current worksheet.

	D	E	F	G	H	I	J	K
1	3	2	1	6				
2	m3	m4	m5	b			total diff=	53
3	-0.00513	-0.00051	-0.00074	0.535574025				
4	-0.00513	-0.00051	-0.00074	0.535574025				
5	gene3	gene4	gene5	5-year survival	m1x1+m2x2+	P(1)	Outcome	Difference
6	350.9466	11.1876	77.926	0	-0.02392719	0.494018	0	0
7	2.579007	301.8684	171.9968	0	0.298006125	0.573955	1	1
8	39.90712	9.2879	104.2632	1	0.373754353	0.592366	1	0
9	2.420307	14.1677	94.6316	1	0.503845363	0.623363	1	0
10	5.831751	35.4298	71.0588	0	0.374075115	0.592443	1	1
11	4.406125	50.444	104.7838	0	0.296268835	0.57353	1	1
12	28.72168	6.4724	78.3981	0	0.374183693	0.59247	1	1
13	1.019015	24.3784	146.7328	1	0.410278594	0.601155	1	0

Figure 6-4. *Examine the temporary classification result*

If your worksheet has a result different from what is shown in Figure 6-4, double check the formulas, especially those in cells H6:K6. Figure 6-5 displays some formulas entered so far. Make sure that yours are the same as those in Figure 6-5.

	H	I	J	K
1				
2			total Diff =	=SUM(K6:K92)
3				
4				
5	m1x1+m2x2+...+b	P(1)	Outcome	Difference
6	=$G$4+SUMPRODUCT($B$4:$F$4, B6:F6)	=1/(1+EXP(0-H6))	=IF(I6<=0.5,0,1)	=IF(G6=J6,0,1)
7	=$G$4+SUMPRODUCT($B$4:$F$4, B7:F7)	=1/(1+EXP(0-H7))	=IF(I7<=0.5,0,1)	=IF(G7=J7,0,1)
8	=$G$4+SUMPRODUCT($B$4:$F$4, B8:F8)	=1/(1+EXP(0-H8))	=IF(I8<=0.5,0,1)	=IF(G8=J8,0,1)
9	=$G$4+SUMPRODUCT($B$4:$F$4, B9:F9)	=1/(1+EXP(0-H9))	=IF(I9<=0.5,0,1)	=IF(G9=J9,0,1)
10	=$G$4+SUMPRODUCT($B$4:$F$4, B10:F10)	=1/(1+EXP(0-H10))	=IF(I10<=0.5,0,1)	=IF(G10=J10,0,1)
11	=$G$4+SUMPRODUCT($B$4:$F$4, B11:F11)	=1/(1+EXP(0-H11))	=IF(I11<=0.5,0,1)	=IF(G11=J11,0,1)
12	=$G$4+SUMPRODUCT($B$4:$F$4, B12:F12)	=1/(1+EXP(0-H12))	=IF(I12<=0.5,0,1)	=IF(G12=J12,0,1)
13	=$G$4+SUMPRODUCT($B$4:$F$4, B13:F13)	=1/(1+EXP(0-H13))	=IF(I13<=0.5,0,1)	=IF(G13=J13,0,1)

Figure 6-5. *Examine the formulas entered so far*

12.   Continue by entering "Likelihood" in cell L5.

13. Enter =IF(G6=1,I6,1-I6) in cell L6. This formula asserts that if the actual outcome is 1, then the likelihood is I6; otherwise, the likelihood is 1-I6.

14. Autofill from cell L6 to cell L92.

15. We need to treat all the likelihood values as a whole. There can be different ways to do it, such as summing all the values in L6:L92, or computing the product of L6:L92. What I am introducing you here is the sum of the logarithm of each likelihood value. Enter the text "Ln(Likelihood)" in cell M5.

16. Enter the formula =IF(L6=0, -1000000, LN(L6)) in cell M6, then autofill from cell M6 to cell M92. Here, the predefined minimum is -1000000 when the likelihood is 0.

17. In cell L1, enter "To-Maximize".

18. Enter the formula =SUM(M6:M92) in cell M1. Cell M1 stores the value to be maximized by Solver.

Part of our worksheet should look like Figure 6-6.

I	J	K	L	M
			To-Maximize	-63.03342831
	total diff=	53		
P(1)	Outcome	Difference	Likelihood	Ln(Likelihood)
0.494018	0	0	0.505981513	-0.681255147
0.573955	1	1	0.426044975	-0.853210364
0.592366	1	0	0.59236585	-0.523630845
0.623363	1	0	0.623362579	-0.47262694
0.592443	1	1	0.407556698	-0.897575219
0.57353	1	1	0.42646985	-0.852213607
0.59247	1	1	0.407530482	-0.897639547

Figure 6-6. *Likelihood computed*

19. The next step is to maximize the value inside cell M1 by using Solver. Select cell M1 ➤ click the main tab Data ➤ click Solver.

20. A menu shows up for us to select proper cells and set up requirements. Follow Figure 6-7 to

   a. Select cell $M$1 for "Set Objective".

   b. Select $B$4:$G$4 for "By Changing Variable Cells".

   c. Choose "Max" as shown in Figure 6-7.

   d. Select the solving method "GRG Nonlinear". Observe that the check box for "Make Unconstrained Variables Non-Negative" is not checked.

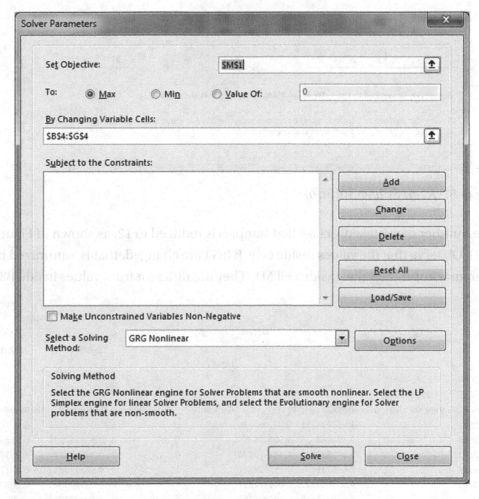

Figure 6-7. *Using Solver to maximize the likelihood*

21. Click the button Solve. Another menu shows up on which make
     sure that "Keep Solver Solution" is chosen as shown in Figure 6-8.

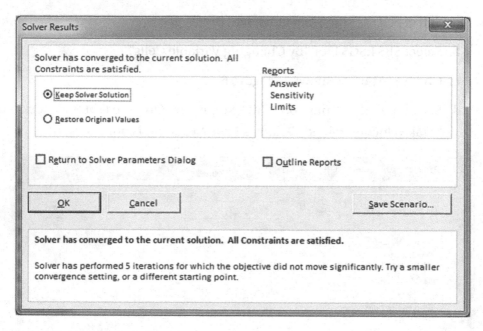

Figure 6-8. *Keep Solver Solution*

The number of mistakenly classified samples is reduced to 12, as shown in Figure 6-9
(cell K2). Observe that the values inside cells B4:G4 are changed, that is, optimized by
Solver to maximize the value inside cell M1. They are different from values inside B3:G3.

	F	G	H	I	J	K	L	M
1	1	6					To-Maximize	-28.72108934
2	m5	b			total Diff =	12		
3	-0.00074	0.535574025						
4	-0.0038	4.222998782						
5	Gene5	5-year survival	m1x1+m2x2+...+b	P(1)	Outcome	Difference	Likelihood	Ln(likelihood)
6	77.926	0	-35.67431731	3.21248E-16	0	0	1	-3.33067E-16
7	171.9968	0	-2.125848326	0.106609769	0	0	0.893390231	-0.112731805
8	104.2632	1	-0.642953181	0.344579274	0	1	0.344579274	-1.065431101
9	94.6316	1	2.049299492	0.885876817	1	0	0.885876817	-0.121177371
10	71.0588	0	-4.987849122	0.006774117	0	0	0.993225883	-0.006797165
11	104.7838	0	-6.928861519	0.000978157	0	0	0.999021843	-0.000978636
12	78.3981	0	0.392794501	0.596955237	1	1	0.403044763	-0.908707649

Figure 6-9. *The result of logistic regression by means of maximizing log likelihoods*

Recall at step 15, I mentioned that we can compute the value in cell M1 differently. For example, we can enter the formula =SUM(L6:L92) or =PRODUCT(L6:L92) in cell M1, or substitute LOG function for LN function in M6:M92, and then use Solver to maximize the value in M1 by optimizing the coefficients in cells B4:G4. I would like to challenge you to try these formulas and compare your results with the method used in this book. You can also try other methods, too.

The aforementioned process is stored in the file chapter6-1b.xlsx. Chapter6-1b.xlsx also includes some scoring data in A1:J101. Please take a look so that you are confident how to apply the constructed logistic regression model to scoring data. Again, the model is just a set of parameters. In this specific case, the set of parameters are in B4:G4.

By Means of Minimizing Log Losses

"Loss function" is a phrase appears very often in machine learning. What is it? Well, we have used it before in linear regression. In linear regression, the least square method is used to minimize the prediction error, commonly called mean square error or quadratic loss, which is the sum of the squares of the differences between predictions and actual values (see Equation 2-1). Here, Equation 2-1 is a bona fide loss function. Thus, we can tell that a loss function is just a mathematical function that quantifies the error between the predicted and actual values in a machine learning method. Indeed, we can understand "loss" as "error" in this sense.

Generally speaking, there are two types of losses in machine learning: regression loss and classification loss. Equation 2-1 is a type of regression loss. The "total diff", that is, the number of mistakenly classified samples in cell K2 which we used earlier in step 11, is a type of classification loss.

A loss function usually works with an optimization function which helps a machine learning model to minimize the loss. In this book, we won't explain optimization function, but we have been using one optimization function employed by the GRG Nonlinear algorithm of Excel Solver to maximize a target value. However, when using a loss function, we want to minimize the target value.

Let's open the file Chapter6-2a.xlsx, you shall notice that the work in this worksheet is near completion as shown in Figure 6-10. Yes, as explained before, log loss is the negation of log of likelihood. So, we do not need to repeat the steps up to the calculation of likelihoods but restart from step 16 to compute the log loss for each likelihood.

▲	A	B	C	D	E	F	G	H	I	J	K	L	M
1		5	4	3	2	1	6						
2		m1	m2	m3	m4	m5	b			total Diff =	53		
3		-0.00032	0.00301	-0.00513	-0.00051	-0.00074	0.535574025						
4		-0.00032	0.00301	-0.00513	-0.00051	-0.00074	0.535574025						
5	PatientID	Gene1	Gene2	Gene3	Gene4	Gene5	5-year survival	m1x1+m2x2+...+b	P(1)	Outcome	Difference	Likelihood	Log loss
6	1	92.0826	443.3735	350.9466	11.1876	77.926	0	-0.023927192	0.494018487	0	0	0.505981513	
7	2	97.1228	29.21562	2.579007	301.8684	171.9968	0	0.298006125	0.573955025	1	1	0.426044975	
8	3	7.73995	42.36842	39.90712	9.2879	104.2632	1	0.373754353	0.59236585	1	0	0.59236585	
9	4	60.2125	25.63164	2.420307	14.1677	94.6316	1	0.503845363	0.623362579	1	0	0.623362579	
10	5	385.4766	20.32964	5.831751	35.4298	71.0588	0	0.374075115	0.592443302	1	1	0.407556698	
11	6	476.644	12.48745	4.406125	50.444	104.7838	0	0.296268835	0.57353015	1	1	0.42646985	
12	7	25.89	18.49515	28.72168	6.4724	78.3981	0	0.374183693	0.592469518	1	1	0.407530482	
13	8	142.8572	15.41199	1.019015	24.3784	146.7328	1	0.410278594	0.601154678	1	0	0.601154678	

Figure 6-10. *The starting point of computing the log loss*

1. Enter the formula =IF(L6=0, 1000000, -LN(L6)) in cell M6, then autofill from cell M6 to cell M92. Here, the predefined maximum loss is 1000000 when the likelihood is 0.

2. In cell L1, enter "To-Minimize".

3. Enter the formula =SUM(M6:M92) in cell M1. Cell M1 stores the value to be minimized by Solver.

4. The next step is to minimize the value inside cell M1 by using Solver. Select cell M1 ➤ click the main tab Data ➤ click Solver.

5. A menu shows up for us to select proper cells and set up requirements. Follow Figure 6-11 to

   a. Select cell $M$1 for "Set Objective".

   b. Select $B$4:$G$4 for "By Changing Variable Cells".

   c. Choose "Min" as shown in Figure 6-11.

   d. Select the solving method "GRG Nonlinear".

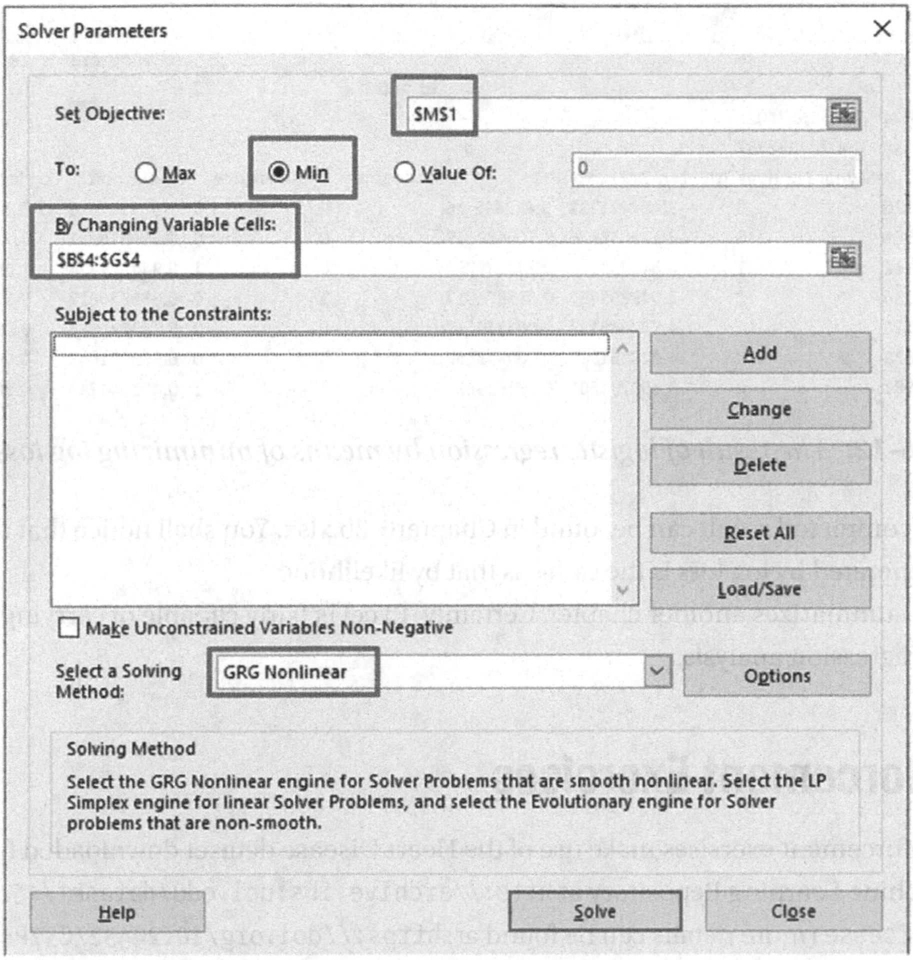

Figure 6-11. Using Solver to minimize the log loss

6. Click the button Solve. Another menu shows up on which make
    sure that "Keep Solver Solution" is chosen.

Part of our worksheet should look like Figure 6-12.

◢	F	G	H	I	J	K	L	M
1	1	6					To-Minimize	28.72108934
2	m5	b			total Diff =	12		
3	-0.00074	0.535574025						
4	-0.0038	4.222998782						
5	Gene5	5-year survival	m1x1+m2x2+…+b	P(1)	Outcome	Difference	Likelihood	Log loss
6	77.926	0	-35.67431731	3.21248E-16	0	0	1	3.33067E-16
7	171.9968	0	-2.125848326	0.106609769	0	0	0.893390231	0.112731805
8	104.2632	1	-0.642953181	0.344579274	0	1	0.344579274	1.065431101
9	94.6316	1	2.049299492	0.885876817	1	0	0.885876817	0.121177371
10	71.0588	0	-4.987849122	0.006774117	0	0	0.993225883	0.006797165
11	104.7838	0	-6.928861519	0.000978157	0	0	0.999021843	0.000978636
12	78.3981	0	0.392794501	0.596955237	1	1	0.403044763	0.908707649

Figure 6-12. *The result of logistic regression by means of minimizing log losses*

The completed result can be found in Chapter6-2b.xlsx. You shall notice that the result generated by log loss is the same as that by likelihood.

This summarizes another chapter. Certainly, Excel is fairly capable of carrying out logistic regression analysis.

Reinforcement Exercises

The reinforcement exercises make use of the Heart Disease dataset downloaded from UCI Machine Learning Repository at http://archive.ics.uci.edu/dataset/45/heart+disease (more details can be found at: https://doi.org/10.24432/C52P4X).

- Chapter6-HW-1.xlsx is for practicing the logistic regression process. Chapter6-HW-1-withAnswers.xlsx presents two solutions: one by maximizing log likelihood and one by minimizing log loss.

- Chapter6-HW-2-cross-validation.xlsx provides an additional exercise for practicing cross-validation as we have just learned cross-validation.

- Chapter6-HW-2-cross-validation-withAnswers.xlsx presents a reference solution for the cross-validation exercise.

Review Points

1. The mechanism of logistic regression

2. Logistic function, sigmoid function, and odds

3. The assumption that log of odds is a linear function

4. The concept of likelihood and log loss

5. Excel functions IF, INDEX, LINEST, SUMPRODUCT, EXP, and LN

6. The use of Solver

CHAPTER 7

K-Nearest Neighbors

Please download the sample Excel files from https://github.com/hhohho/Learn-Data-Mining-through-Excel-2 for this chapter's exercises.

General Understanding

People tend to make decisions or take actions based on the advice from people around them. It happens from time to time in our lives that we take suggestions from people around us, especially from those who are very close to us. The data mining method, k-nearest neighbor (K-NN) method, is a reflection of such a real-life moment. Here, k indicates that the decision is based on k neighbors. For example, among 10000 known samples, when a new case A happens and we need to predict the outcome of A, we find 11 (usually odd) nearest neighbors of A from the samples, and based on the majority outcome of the 11 neighbors (k = 11 in this case), the outcome of A is predicted. In K-NN, a neighbor with a certain outcome is usually expressed as a vote to the outcome.

In Chapter 1, I mentioned to consider data mining and machine learning the same in this book. When talking about machine learning, there are two types of learners (methods): eager learners and lazy learners. LDA and logistic regression are both eager learners because they construct a model based on the training dataset first, then apply the model to any future data. Such a model is fundamentally a parameter set or rule set. On the contrary, K-NN is a lazy learner. A lazy learner does not construct a model first; it does not generate a set of parameters or rules based on the training dataset. Instead, when scoring data come in, a lazy learner makes use of the whole set of the training data to dynamically classify the scoring data. In the case of K-NN, the nearest neighbors of a scoring data point are dynamically computed using the whole training dataset. Since there is no parameter set precalculated, K-NN is also a nonparametric data mining method.

© Hong Zhou 2023
H. Zhou, *Learn Data Mining Through Excel*, https://doi.org/10.1007/978-1-4842-9771-1_7

The major work in K-NN is to determine the neighbors that are nearest. Given a scoring data point A, what can be defined as its nearest neighbors? In another words, how do we measure proximity?

One way to measure proximity is by calculating the distance between two data points. When we discuss k-means clustering (be aware to distinguish k-means clustering and K-NN), we make use of Euclidean distance. Certainly, we can do the same here. In fact, Euclidean distance is the most used proximity metric in K-NN.

Manhattan and Chebyshev distances are sometimes used to measure proximity, too. Given two data points (x1, x2, x3) and (y1, y2, y3). The Manhattan distance between the two data points is computed as $|y1 - x1| + |y2 - x2| + |y3 - x3|$.

Between two data points, the Chebyshev distance is the maximum difference between all attributes. For example, given two data points (1, 2, 3) and (4, 3, 7). The three differences are $|4 - 1| = 3$, $|3 - 2| = 1$, and $|7 - 3| = 4$. The Chebyshev distance is 4 since 4 is the maximum.

Euclidean, Manhattan, and Chebyshev distances are for numerical data points. For discrete attributes, Hamming distance can be used. Given two strings of the same length, Hamming distance counts the number of positions where the two corresponding symbols are different. For example, given two binary strings 10011010 and 11001011, their Hamming distance is 3. This is shown as follows:

    1**0**0**1**101**0**

    1**1**0**0**101**1**

There are a few other methods to measure proximity. In this chapter, we are going to use Euclidean distance again.

Learn K-NN Through Excel

Chapter7-1a.xlsx simulates a business scenario. A credit card company starts a marketing campaign. The company wants to predict if a given new customer Daisy will accept a specific credit card offer based on how other similar customers have responded. Suppose there are already 200 customers who have responded, and it is believed that the three attributes, age, income, and number of credit cards owned, are the crucial factors determining the likelihood of acceptance.

Experiment 1

Open the file Chapter7-1a.xlsx. Column E stores the responses of the existing contacted people. Number 1 indicates offer acceptance, while 0 represents offer denial. As shown in Figure 7-1, cells G1:J9 are highlighted where we need to find Daisy's likely responses based on different k values.

◢	A	B	C	D	E	F	G	H	I	J
1	Name	Age	Income (1000s)	Cards have	Response			Age	Income	Cards have
2	N1	71	32	3	0		**Daisy**	27	155	5
3	N2	33	144	8	1		K	Daisy's likely reponse:		
4	N3	49	63	10	0		1			
5	N4	38	57	10	0		3			
6	N5	26	159	5	0		5			
7	N6	30	163	8	1		7			
8	N7	35	41	0	0		9			
9	N8	55	44	9	1		11			
10	N9	60	10	3	0					

Figure 7-1. *A glance of the data for K-NN data mining*

The first job is to determine how to compute the distance between Daisy and any other persons based on the three attributes: Age, Income, and Cards have. Let's utilize Euclidean distance again. Follow these instructions to complete Experiment 1:

1. Enter "Distance" in cell F1. In cell F2, enter the following formula:

    =SQRT(SUMXMY2(B2:D2, $H$2:$J$2))

2. Autofill from F2 to F201. Part of our worksheet looks like Figure 7-2.

▲	A	B	C	D	E	F	G	H	I	J
1	Name	Age	Income (1000s)	Cards have	Response	Distance		Age	Income	Cards have
2	N1	71	32	3	0	130.648	**Daisy**	27	155	5
3	N2	33	144	8	1	12.8841	K	Daisy's likely reponse:		
4	N3	49	63	10	0	94.7259	1			
5	N4	38	57	10	0	98.7421	3			
6	N5	26	159	5	0	4.12311	5			
7	N6	30	163	8	1	9.05539	7			
8	N7	35	41	0	0	114.39	9			
9	N8	55	44	9	1	114.547	11			
10	N9	60	10	3	0	148.721				

Figure 7-2. *K-NN data to predict Daisy's response*

3. I will make use of the function SMALL to find the nearest neighbors based on the given K values. Enter "Small" in cell K3 and the following formula in cell K4:

=SMALL(F$2:F$201,G4)

The function SMALL returns the kth smallest value in an array. As G4 = 1, the preceding formula finds the first smallest value in F2:F201.

4. Autofill from cell K4 to cell K9.

Using k=3 as an example (corresponding to cell G5): since G5 = 3, the formula in cell K5 finds the third smallest value in F2:F201. Any data points whose distance to Daisy is smaller than or equal to the value in cell K5 are a nearest neighbor to Daisy.

5. Enter number 1 in cell L3 and number 0 in cell M3. Column L records the number of response 1, while column M records the number of response 0.

6. Enter the following formula in cell L4:

=COUNTIFS($E$2:$E$201,L$3,$F$2:$F$201,"<="&$K4)

Note, the expression "<="&$K4 is translated to "<=4.12311". The operator & concatenates the text "<=" with the value in cell K4 (which is 4.12311).

7. Autofill from cell L4 to M4, and then autofill together to L9:M9.

The current result is illustrated in Figure 7-3.

▲	G	H	I	J	K	L	M
1		Age	Income	Cards have			
2	**Daisy**	27	155	5			
3	K	Daisy's likely reponse:			Small	1	0
4	1				4.12311	0	1
5	3				8.12404	1	2
6	5				9.05539	3	2
7	7				11.4891	4	3
8	9				12.3288	5	4
9	11				13.1529	7	4

Figure 7-3. *Compute the votes for 1 and 0*

8. Based on the numbers of 1s and 0s, that is, the number of votes for acceptance and denial respectively, Daisy's response can be determined by the following formula which should be entered in cell H4 (note that cells H4, I4, and J4 are merged, and therefore H4 represents all the three cells):

    =IF(L4>M4,$L$3,$M$3)

9. Autofill from cell H4 to cell H9.

Figure 7-4 presents the final result. The result shows that with a different k value, the prediction of Daisy's response based on the majority votes from her "neighbors" would be different. Be aware that K-NN is a probability model, which means it computes the probability of how Daisy will respond based on her neighbors' former responses. For example, when k equals 5 (5 is a common choice for K), K-NN predicts that she is likely to accept the credit card offer. Her probability of acceptance is 3/5. Still, she has 2/5 probability to reject the credit card offer. Observe when k=1, her rejection probability is 100%. This does not guarantee that she will not accept the credit card offer, however.

◢	G	H	I	J	K	L	M
1		Age	Income	Cards have			
2	**Daisy**	27	155	5			
3	K	Daisy's likely reponse:			Small	1	0
4	1		0		4.12311	0	1
5	3		0		8.12404	1	2
6	5		1		9.05539	3	2
7	7		1		11.4891	4	3
8	9		1		12.3288	5	4
9	11		1		13.1529	7	4

Figure 7-4. *K-NN predicts Daisy's responses based on different K values*

It is not difficult to notice that the K value matters in the K-NN model. Which K value is the best is debatable, but the general understanding is that K should better be an odd integer.

If the data is large enough for cross-validation, then we can compute the cross-validation errors with different K values and plot the errors against K values. The plot should look like L or U shape, that is, it has an elbow. Usually the optimal K value is at the elbow (if it is U shape, the first elbow). This is the so-called Elbow method or Elbow Curve Validation technique for finding an optimal K value for K-NN.

Conducting the Elbow Curve Validation through Excel is doable but not practical, so I won't show how to implement it here.

Experiment 2

In real life, we often take the suggestions more seriously if they are from whom we trust the most. So, should we treat Daisy's neighbors differently since they can have different distances? Yes, we should. For example, we should weigh more heavily the votes from the neighbors closer to her.

The fundamental concept of K-NN is that data points near to each other are alike to each other and belong to the same class. The closer two data points are, the more likely they belong to the same class. In contrast, the further two data points are, the more likely they belong to different classes. When assigning weights to neighbors, a rule to follow is that the larger the distance, the less weight of the vote.

Which function can incorporate the distance factor into the vote properly? A simple solution is to use the reciprocal of the distance. However, using reciprocal can cause an error when one neighbor's distance to Daisy is 0. To avoid such an error, a modified reciprocal function can be used:

$$w = \frac{1}{1+bd} \tag{7-1}$$

where b is a constant and should better be greater than or equal to 1 and d is the Euclidean distance between two data points.

A more common function takes the form of the following equation:

$$w = \frac{1}{a^d} \tag{7-2}$$

where $a > 1$. Often a is set to e, the base of the natural logarithm. Note, w decreases exponentially along with an increasing d, especially when a = e, in which case a closer data point can easily become the dominant vote.

In this experiment, we adopt Equation (7-1) and set b = 1. Open Chapter7-2a.xlsx. The setup of data in this file looks like Figure 7-5.

◢	A	B	C	D	E	F	G
1		Age	Income	Cards have			
2	**Daisy**	27	155	5			
3	K	Daisy's likely reponse:					
4	1						
5	3						
6	5						
7	7						
8	9						
9	11						
10							
11	Name	Age	Income (1000s)	Cards	Response		
12	N1	71	32	3	0		
13	N2	33	144	8	1		
14	N3	49	63	10	0		
15	N4	38	57	10	0		

Figure 7-5. *Data setup for weighted K-NN analysis*

Follow these instructions to complete the K-NN analysis with weighted votes:

1. Enter "Distance" in cell F11. The distance to Daisy from each neighbor is computed in column F. This can be achieved by entering the following formula in cell F12 and then autofill from F12 to F211:

   ```
   =SQRT(SUMXMY2(B12:D12, B$2:D$2))
   ```

2. In cells G11 and H11, enter "Reciprocal" and "Weight", respectively.

3. In cell G12, enter the formula `=1/(1+F12)`, and then autofill from G12 to G211.

4. In cell H12, enter the formula `=G12/SUM($G$12:$G$211)`. This formula guarantees that the sum of weights in column H is 1.

5. Autofill from H12 to H211.

110

By now, the weights of all neighbor's votes have been computed, and they are inversely proportional to the distance between Daisy and each neighbor. Currently, part of our worksheet should look like Figure 7-6.

▲	A	B	C	D	E	F	G	H
1		Age	Income	Cards have				
2	**Daisy**	27	155	5				
3	K	Daisy's likely reponse:						
4	1							
5	3							
6	5							
7	7							
8	9							
9	11							
10								
11	Name	Age	Income (1000s)	Cards	Response	Distance	Reciprocal	Weight
12	N1	71	32	3	0	130.64838	0.007596	0.0015689
13	N2	33	144	8	1	12.884099	0.0720248	0.014876
14	N3	49	63	10	0	94.72592	0.0104465	0.0021576
15	N4	38	57	10	0	98.742088	0.0100259	0.0020707

Figure 7-6. *Compute the weight for each neighbors' vote*

6. In cells E3, F3, G3, and H3, enter "Small", "1", "0", and "Probability", respectively, as shown in Figure 7-7.

7. In cell E4, enter the formula =SMALL(F$12:F$211,A4) and autofill from E4 to E9.

8. In cell F4, enter the following formula:

   =SUMIFS($H$12:$H$211,$E$12:$E$211, F$3,$F$12:$F$211,"<="&$E4)

9. Autofill from F4 to G4 and then F4 to F9 and G4 to G9.

10. In cell H4, enter the formula =F4/SUM(F4:G4). The value in cell H4 is the acceptance probability when k=1.

11. Autofill from H4 to H9.

By now, part of our worksheet should look like Figure 7-7.

▲	A	B	C	D	E	F	G	H
1		Age	Income	Cards have				
2	**Daisy**	27	155	5				
3	K	Daisy's likely reponse:			Small	1	0	Probability
4	1				4.12311	0	0.0403154	0
5	3				8.12404	0.0226369	0.0788569	0.2230373
6	5				9.05539	0.0646866	0.0788569	0.4506408
7	7				11.4891	0.0834629	0.0953945	0.4666449
8	9				12.3288	0.0994535	0.1108903	0.4728141
9	11				13.1529	0.128923	0.1108903	0.5375973
10								
11	Name	Age	Income (1000s)	Cards	Response	Distance	Reciprocal	Weight
12	N1	71	32	3	0	130.64838	0.007596	0.0015689

Figure 7-7. *Compute the weighted votes*

12. Now it is time to classify Daisy's responses. Enter the formula
     =IF(H4>0.5,"Yes","No") inside cell B4.

13. Autofill from B4 to B9.

Our result should look like Figure 7-8. Certainly, with the weighted votes in the K-NN model, we are more confident that Daisy is likely to reject the credit card offer than we were before. The complete result is available in the file Chapter7-2b.xlsx.

▲	A	B	C	D	E	F	G	H
1		Age	Income	Cards have				
2	**Daisy**	27	155	5				
3	K	Daisy's likely reponse:			**Small**	1	0	**Probability**
4	1		No		4.12311	0	0.0403154	0
5	3		No		8.12404	0.0226369	0.0788569	0.2230373
6	5		No		9.05539	0.0646866	0.0788569	0.4506408
7	7		No		11.4891	0.0834629	0.0953945	0.4666449
8	9		No		12.3288	0.0994535	0.1108903	0.4728141
9	11		Yes		13.1529	0.128923	0.1108903	0.5375973
10								
11	Name	Age	Income (1000s)	Cards	Response	Distance	Reciprocal	Weight
12	N1	71	32	3	0	130.64838	0.007596	0.0015689
13	N2	33	144	8	1	12.884099	0.0720248	0.014876

Figure 7-8. *Applying weighted votes in the K-NN model*

In this experiment, the cutoff we used here is 0.5, that is, if the acceptance probability is greater than 0.5, we think Daisy's response is likely to be Yes, otherwise No. Assigning 0.5 as the cutoff is a natural decision. However, when assessing our K-NN model by cross-validation, it would be interesting to pick different cutoff values and conduct a ROC analysis.

Can we improve our weighted K-NN model even more? The answer might be yes. It is not difficult to notice that attribute Income contributes the most to the distances, while the attribute "Cards have" contributes the least. We understand that different attributes can have different impacts on determining Daisy's response, which is part of the business understanding phase in developing a data mining model. However, when we are not clear which attribute contributes more, it is better to treat all attributes the same. Therefore, it is necessary to normalize all the three attributes such that they are on the same scale, say 0–100.

Experiment 3

Normalization can be done in different ways. Z-score normalization is based on the data's variance. In this experiment, we perform a range normalization, that is, all attributes are normalized based on their respective maximum and minimum values.

We would like the range to be 0–100. Thus, the minimum value and maximum value are transformed to 0 and 100, respectively, for each attribute.

Let's follow these step-by-step instructions to exercise this experiment:

1. Open Chapter7-3a.xlsx. In cells J1:J6 and K1:K6, enter the texts or formulas as shown in Figure 7-9. These values are the actual ranges of the three attributes.

◢	J	K
1	max-age	=MAX(B12:B211,B2)
2	min-age	=MIN(B12:B211,B2)
3	max-income	=MAX(C12:C211, C2)
4	min-income	=MIN(C12:C211,C2)
5	max-cards	=MAX(D12:D211,D2)
6	min-cards	=MIN(D12:D211,D2)

Figure 7-9. *Define the ranges*

The expression to normalize a value is 100 / (max – min) * (value – min). Note for a given attribute, if its max equals its min, this expression can fail. However, such a case is very unlikely. When such a case occurs, this attribute has no variance and therefore has no value for our data mining project. I will discuss this in a later chapter when I introduce exploratory data analysis (EDA).

2. In cells F11, G11, and H11, enter "normalized age", "normalized income", and "normalized cards", respectively.

3. In cell F12, enter the following formula and then autofill from F12 to F211:

```
=100/($K$1-$K$2)*(B12-$K$2)
```

4. In cell G12, enter the following formula and then autofill from G12 to G211:

```
=100/($K$3-$K$4)*(C12-$K$4)
```

5. In cell H12, enter the following formula and then autofill from H12 to H211:

```
=100/($K$5-$K$6)*(D12-$K$6)
```

Part of our worksheet should look like Figure 7-10.

	A	B	C	D	E	F	G	H	I	J	K
1		Age	Income	Cards have						max-age	75
2	Daisy	27	155	5						min-age	18
3	K	Daisy's likely reponse:								max-income	200
4	1									min-income	10
5	3									max-cards	10
6	5									min-cards	0
7	7										
8	9										
9	11										
10											
11	Name	Age	Income (1000s)	Cards	Response	normalized age	normalized income	normalized cards			
12	N1	71	32	3	0	92.9824561	11.5789474	30			
13	N2	33	144	8	1	26.3157895	70.5263158	80			
14	N3	49	63	10	0	54.3859649	27.8947368	100			
15	N4	38	57	10	0	35.0877193	24.7368421	100			

Figure 7-10. *Normalize all attributes*

Continue the following steps as we did before to accomplish this experiment:

6. In cell E2, enter the formula =100/(K1-K2)*(B2-K2). This is the normalized "Age" of Daisy.

7. In cell F2, enter the formula =100/(K3-K4)*(C2-K4). This is the normalized "Income" of Daisy.

8. In cell G2, enter the formula =100/(K5-K6)*(D2-K6). This is the normalized "Cards have" of Daisy.

9. Enter "distance", "reciprocal", and "weight" in cells I11, J11, and K11, respectively.

10. Enter "Small", "1", "0", and "Probability" in cells E3, F3, G3, and H3, respectively.

Part of our worksheet now looks like Figure 7-11.

E	F	G	H	I	J	K
					max-age	75
15.78947	76.3157895	50			min-age	18
Small	1	0	Probability		max-income	200
					min-income	10
					max-cards	10
					min-cards	0
Response	normalized age	normalized income	normalized cards	distance	reciprocal	weight

Figure 7-11. *Setup of experiment 3*

11. In cell I12, enter the following formula and autofill from I12 to I211:

    `=SQRT(SUMXMY2(E$2:G$2, F12:H12))`

12. In cell J12, enter the formula `=1/(1+I12)` and autofill from J12 to J211.

13. In cell K12, enter the formula `=J12/SUM($J$12:$J$211)` and autofill from K12 to K211.

Make sure our worksheet looks like Figure 7-12. If it is different, we'd better examine our formulas.

	F	G	H	I	J	K
11	normalized age	normalized income	normalized cards	distance	reciprocal	weight
12	92.982456	11.57895	30	102.71132	0.009642149	0.002483575
13	26.315789	70.52632	80	32.31596091	0.030015643	0.007731274
14	54.385965	27.89474	100	79.58823703	0.012408759	0.003196184
15	35.087719	24.73684	100	74.38286157	0.013265615	0.003416888
16	14.035088	78.42105	50	2.740438483	0.267348335	0.068862198
17	21.052632	80.52632	80	30.74783509	0.031498211	0.008113146

Figure 7-12. *Compute weights after normalization*

Continue the task by following these instructions:

14. Enter the formula =SMALL(I$12:I$211,A4) inside cell E4. Autofill to E9.

15. Enter the following formula in cell F4:

    =SUMIFS($K$12:$K$211,$E$12:$E$211, F$3,$I$12:$I$211,"<="&$E4)

16. Autofill from cell F4 to G4 and from F4 to F9 and G4 to G9.

17. Enter the formula =F4/SUM(F4:G4) in cell H4, then autofill from H4 to H9.

18. Enter the formula =IF(H4>0.5,"Yes","No") in cell B4, then autofill from B4 to B9.

The final result is shown in Figure 7-13. Clearly, when the data are processed differently, the results can be different (compare Figure 7-13 with Figure 7-8 and Figure 7-4). It is always a good idea to understand how our data are processed when applying the K-NN method. The complete result is available in Chapter7-3b.xlsx.

◢	B	C	D	E	F	G	H	I	J	K
1	Age	Income	Cards have						max-age	75
2	27	155	5	15.78947	76.3157895	50			min-age	18
3	Daisy's likely reponse:			Small	1	0	Probability		max-income	200
4		No		2.740438	0	0.0688622	0		min-income	10
5		No		13.53267	0.03860176	0.0688622	0.35920657		max-cards	10
6		No		21.34303	0.05012996	0.08135458	0.38126124		min-cards	0
7		No		22.89995	0.07178676	0.08135458	0.46876146			
8		No		23.37287	0.07178676	0.10263324	0.41157411			
9		No		25.1006	0.09209807	0.10263324	0.47294947			
10										
11	Age	Income (1000s)	Cards	Response	normalized age	normalized income	normalized cards	distance	reciprocal	weight
12	71	32	3	0	92.9824561	11.5789474	30	102.7113	0.009642149	0.002484
13	33	144	8	1	26.3157895	70.5263158	80	32.31596	0.030015643	0.007731
14	49	63	10	0	54.3859649	27.8947368	100	79.58824	0.012408759	0.003196
15	38	57	10	0	35.0877193	24.7368421	100	74.38286	0.013265615	0.003417
16	26	159	5	0	14.0350877	78.4210526	50	2.740438	0.267348335	0.068862

Figure 7-13. *A K-NN method with normalized attributes and weighted votes*

Experiment 4

In the previous three experiments, the target variable has only two distinct responses. Let's conduct a K-NN analysis when the target variable has more than two responses. Open Chapter7-4a.xlsx. This file is very similar to Chapter7-3b.xlsx except (1) its responses are Yes, Not Sure, and No; (2) it has not computed the probability for each possible response or Daisy's likely responses. Take a look at the file first. The worksheet looks like Figure 7-14. Note in cell I3, it is "Max Weight", not "Probability".

	A	B	C	D	E	F	G	H	I	J	K
1		Age	Income	Cards have						max-age	75
2	Daisy	27	155	5	15.78947	76.3157895	50			min-age	18
3	K	Daisy's likely reponse:			Small	Yes	Not Sure	No	Max Weight	max-income	200
4	1				2.740438					min-income	10
5	3				10.61077					max-cards	10
6	5				13.53267					min-cards	0
7	7				21.34303						
8	9				22.89995						
9	11				23.37287						
10											
11	Name	Age	Income (1000s)	Cards	Response	normalized age	normalized income	normalized cards	distance	reciprocal	weight
12	N1	71	32	3	No	92.9824561	11.5789474	30	102.71132	0.009642149	0.002365
13	N2	33	144	8	No	26.3157895	70.5263158	80	32.31596091	0.030015643	0.007362
14	N3	29	163	5	Yes	19.2982456	80.5263158	50	5.480876966	0.154300106	0.037847
15	N4	38	57	10	No	35.0877193	24.7368421	100	74.38286157	0.013265615	0.003254
16	N5	26	159	5	Not Sure	14.0350877	78.4210526	50	2.740438483	0.267348335	0.065576
17	N6	30	163	8	No	21.0526316	80.5263158	80	30.74783509	0.031498211	0.007726
18	N7	35	41	0	No	29.8245614	16.3157895	0	79.35353607	0.012445003	0.003053
19	N8	55	44	9	No	64.9122807	17.8947368	90	86.17464569	0.011471225	0.002814
20	N9	60	10	3	Yes	73.6842105	0	30	97.85652905	0.01011567	0.002481

Figure 7-14. K-NN with three distinct categorical outcomes

Follow these instructions to complete the final computation:

1. Enter the following formula in cell F4:

    =SUMIFS($K$12:$K$211,$E$12:$E$211, F$3,$I$12:$I$211,"<="&$E4)

2. Autofill from F4 to H4, then autofill all together to row 9.

3. Enter the formula =MAX(F4:H4) in cell I4, and autofill from I4 to I9.

4. In cell B4, enter the following formula and then autofill to cell B9:

    =INDEX(F$3:H$3,1,MATCH(I4,F4:H4,0))

    Note the use of MATCH function. The last parameter inside MATCH is 0, indicating it is searching for an exact match. This is important here as many numbers are close to each other.

That's it. The final result is shown in Figure 7-15 and can be available in the file Chapter7-4b.xlsx.

◢	A	B	C	D	E	F	G	H	I	J	K
1		Age	Income	Cards have						max-age	75
2	Daisy	27	155	5	15.78947	76.3157895	50			min-age	18
3	K	Daisy's likely reponse:			Small	Yes	Not Sure	No	Max Weight	max-income	200
4	1		Not Sure		2.740438	0	0.06557647	0	0.065576473	min-income	10
5	3		Not Sure		10.61077	0.05897309	0.06557647	0	0.065576473	max-cards	10
6	5		Yes		13.53267	0.09573298	0.06557647	0	0.095732981	min-cards	0
7	7		Yes		21.34303	0.10671112	0.07747279	0	0.106711116		
8	9		Yes		22.89995	0.1169741	0.08783326	0	0.1169741		
9	11		Yes		23.37287	0.1169741	0.09803278	0.01006384	0.1169741		
10											
11	Name	Age	Income (1000s)	Cards	Response	normalized age	normalized income	normalized cards	distance	reciprocal	weight
12	N1	71	32	3	No	92.9824561	11.5789474	30	102.71132	0.009642149	0.002365
13	N2	33	144	8	No	26.3157895	70.5263158	80	32.31596091	0.030015643	0.007362
14	N3	29	163	5	Yes	19.2982456	80.5263158	50	5.480876966	0.154300106	0.037847
15	N4	38	57	10	No	35.0877193	24.7368421	100	74.38286157	0.013265615	0.003254
16	N5	26	159	5	Not Sure	14.0350877	78.4210526	50	2.740438483	0.267348335	0.065576
17	N6	30	163	8	No	21.0526316	80.5263158	80	30.74783509	0.031498211	0.007726
18	N7	35	41	0	No	29.8245614	16.3157895	0	79.35353607	0.012445003	0.003053
19	N8	55	44	9	No	64.9122807	17.8947368	90	86.17464569	0.011471225	0.002814
20	N9	60	10	3	Yes	73.6842105	0	30	97.85652905	0.01011567	0.002481

Figure 7-15. *The final result of K-NN with three predictive outcomes*

Reinforcement Exercises

The reinforcement exercise in Chapter7-HW.xlsx may take you some effort and caution. But it is necessary for you to become more skilled with the K-NN data mining method. Should you run into any difficulty completing this reinforcement exercise, feel free to take a look at the solution inside Chapter7-HW-withAnswers.xlsx.

Review Points

1. What is K-NN and its resemblance in our life

2. Different distance functions

3. Different weight functions and normalization techniques

4. The Elbow method for finding the optimal K value for K-NN

5. Excel functions SQRT and SMALL and Excel operator &

6. Excel functions IF, COUNTIFS, SUMIFS, MATCH, and INDEX

Hierarchical Clustering and Dendrogram

You may wonder why I do not explain hierarchical clustering right after k-means clustering. Well, there are some reasons. The first one is that Excel is not as suitable for hierarchical clustering or dendrogram as it is for k-means clustering. I would like to introduce a few more other popular data mining methods before diving into this challenging topic. The second reason is that I consider it suitable to explain hierarchical clustering right after K-NN. The last reason is because of dendrogram. To draw a dendrogram in Excel requires quite some manual work, which could be boring to some learners.

Please download the sample Excel files from https://github.com/hhohho/Learn-Data-Mining-through-Excel-2 for this chapter's exercises.

General Understanding

Hierarchical clustering is similar to k-means clustering in that it is also an unsupervised classification method that categorizes subjects (data points) into different groups (clusters) based on the similarity of certain characteristics. However, these two clustering methods are based on two different algorithms, which results in different final products. K-means clustering generates parallel clusters. Hierarchical clustering produces a tree of clusters as its name indicates, instead. Moreover, the final product of hierarchical clustering is usually presented in the form of a dendrogram that shows the relative distances of the original data points or grouped data points. This is the major reason why hierarchical clustering is frequently used in biomedical sciences to show the clustering of genes or samples.

K-means clustering starts with defining the number k, the number of clusters. Hierarchical clustering starts with a proximity matrix and never needs to predefine the number of clusters. The algorithm of hierarchical clustering can be explained by the following example that contains six data points A, B, C, D, E, and F.

1. Every data point is treated as a cluster initially.

2. A proximity matrix is constructed to show the distances among the six clusters.

3. Assume that the smallest distance is between A and C, then A and C are placed in the same cluster in the first round and they are represented as AC.

4. Now, there are five clusters only: AC, B, D, E, and F.

5. Repeat step 2 to create a proximity matrix for the five clusters. Assume that B and D have the least distance, then there are only four clusters: AC, BD, E, and F.

6. Repeat step 2 to create a proximity matrix for the four clusters. Assume that AC and F have the least distance, then there are only three clusters: ACF, BD, and E.

7. Repeat step 2 to create a proximity matrix for the three clusters. Assume that ACF and BD have the least distance, then there are only two clusters: ACFBD and E.

8. Hierarchical clustering stops at only one cluster: ACFBDE.

The preceding algorithm is for agglomerative clustering, a bottom-up hierarchical clustering. The opposite of agglomerative clustering is called divisive clustering, which is top down. In this book, we will experience agglomerative clustering, and the distance among clusters is again the Euclidean distance.

What is dendrogram? Well, a graph truly worth more than one thousand words in this case. Following the preceding example, the final product of the algorithm can be represented by the following diagram which is a bona fide dendrogram.

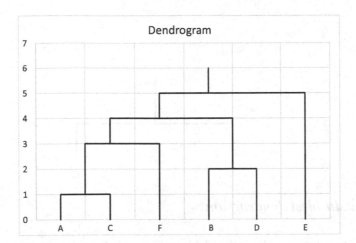

Figure 8-1. *A sample dendrogram showing the hierarchical clustering process of A,B,C,D,E,F*

Observe in Figure 8-1 that (1) there are five levels for six data points, one level for one round of clustering and (2) A and C are first clustered together, and therefore the clade for AC has the least height. In dendrogram, the higher the clade, the larger the distance between the two clusters.

Draw a Dendrogram in Excel Without Add-ins

Before I show you how to use Excel to implement the agglomerative hierarchical clustering algorithm, let's learn how to draw a dendrogram first in Excel. Note, there are commercial analytical software that can be purchased and added into Excel to ease the hierarchical clustering process and dendrogram creation. What I am showing here does not require any add-ins.

Let's start from the simplest dendrogram as shown in Figure 8-2.

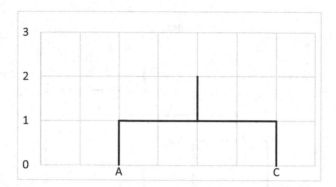

Figure 8-2. *The simplest dendrogram*

If I ask you to use a pen to draw the chart in Figure 8-2 on a piece of paper under the condition that your pen can never leave the paper until completion, how are you going to draw it? Well, you can start from the top, right, or left. Assume you start from the left where the letter A stays (height=0), your pen should move as follows:

1. Start from A, go up by 1.

2. Turn to right, stop at above letter C.

3. Go down by 1 to letter C.

4. Go up by 1.

5. Turn left and stop at the middle between A and C.

6. Go up by 1 to 2.

What if the dendrogram looks like Figure 8-3 and we need to stop at height 3?

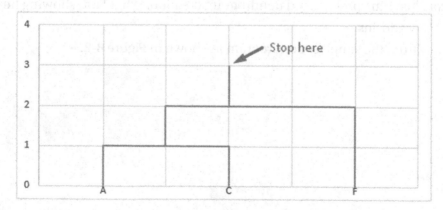

Figure 8-3. *A slightly more complicated dendrogram*

Well, we can continue moving our pen as follows:

7. Turn our pen to right so that it is right above letter F.

8. Go down by 2 to 0.

9. Go up by 2.

10. Turn left and stop above letter C.

11. Go up by 1 to 3.

So far, we should have understood that our pen goes through some paths twice. Yes, that is the trick. If we convert these pen movements into a scatter plot in Excel, then the movements can be represented by the data points shown in Figure 8-4.

◢	A	B	C	D	E	F	G	H
1	X	Y						
2	1	0		Start from (1, 0)				
3	1	1		Go up by 1 (1, 1)				
4	3	1		Go right by 2 (3, 1)				
5	3	0		Go down by 1 (3, 0)				
6	3	1		Go up by 1. Here, go back to (3, 1)				
7	2	1		Go left by 1. Here, go back to (2, 1)				
8	2	2		Go up by 1 to (2, 2)				
9	5	2		Go right by 3 (5, 2)				
10	5	0		Go down by 2 (5, 0)				
11	5	2		Go up by 2. Here, go back to (5, 2)				
12	3	2		Go left by 2 (3, 2)				
13	3	3		Go up by 1 (3, 3)				

Figure 8-4. *The X and Y values for a scatter plot that produces a dendrogram*

Please open Chapter8-1a.xlsx. What will jump into your eyes should look like Figure 8-5, which is a dendrogram slightly different from Figure 8-3 because it does not have the labels A, C, or F. Note, this dendrogram is a scatter plot and is produced based on the data in A1:B13 which is the same as shown in Figure 8-4.

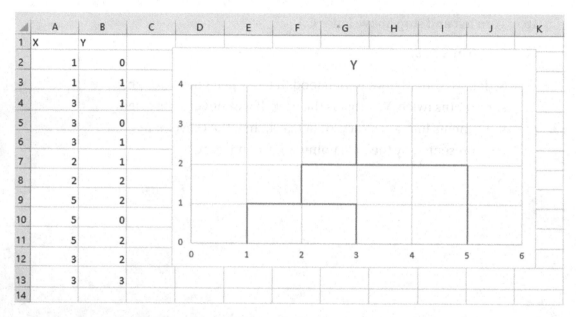

Figure 8-5. *The dendrogram without labels*

To make the dendrogram in Figure 8-5 look like that in Figure 8-3, we need to complete the following:

12. As shown in Figure 8-6, click inside the x-axis, then select Delete on the upcoming menu. This action deletes the x-axis.

13. Right-click on the chart title to delete it, too, as shown in Figure 8-6.

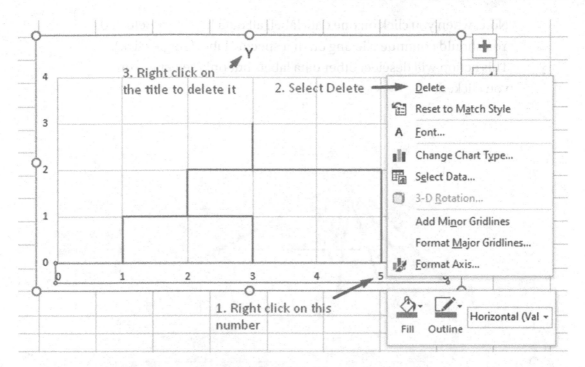

Figure 8-6. *Delete the x-axis and the chart title*

14. Follow Figure 8-7 to add data labels. Make sure to select Below.

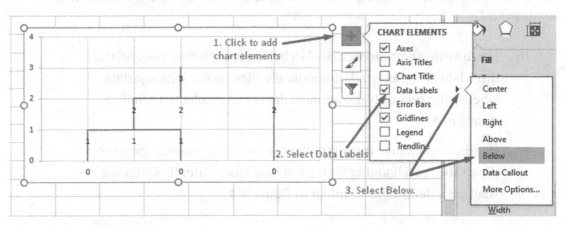

Figure 8-7. *Add data labels*

15. Follow Figure 8-8 to delete data labels one by one but leave the bottom three labels intact. Bear in mind that this action requires some effort.

Note, when you click on one data label, all data labels are selected. You should continue clicking on that specific label (single click). This action will deselect other data labels but only select the one you clicked.

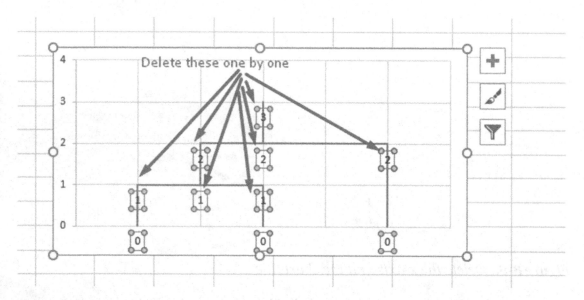

Figure 8-8. *Delete the marked data labels one by one. Leave the bottom three labels intact*

16. For the three bottom data labels, click on the leftmost one. All the three labels are selected. Immediately click on the leftmost data label again; now only the leftmost data label is selected. Change its text to A.

17. Continue to change the text of the rest two data labels. Figure 8-9 shows the text changing for the last data label. After this step, we shall get a dendrogram similar to Figure 8-3.

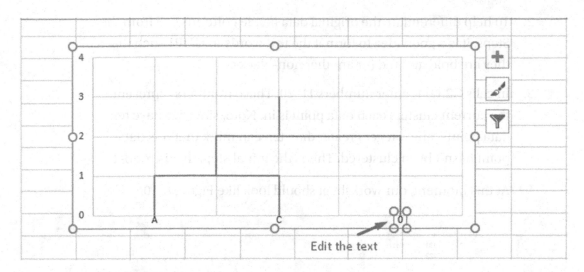

Figure 8-9. *Edit the text inside a data label*

The final result is available in Chapter8-1b.xlsx.

Learn Hierarchical Clustering Through Excel

Let's open the file Chapter8-2a.xlsx which contains ten data records. Using Excel to demonstrate hierarchical clustering with too many records won't be practical because both dendrogram creation and the hierarchical clustering process require quite some manual work. As the purpose of this chapter is to let you experience how hierarchical clustering works, there is no need to start with too many records.

You may have already noticed that these ten records are the first ten records from Chapter3-1a.xlsx. These ten records are for the ten countries L1, L2, ..., L10 (column B). Our task is to hierarchically cluster these ten data points and then observe their relative distances or similarities.

Please follow these instructions to execute the hierarchical clustering algorithm as explained earlier:

1. In cells G1:P1, enter the text words "Step-1", "Step-2", "Step-3", "Step-4", "Step-5", "Step-6", "Step-7", "Step-8", "Step-9", and "Step-10". You can achieve this by entering "Step-1" in cell G1 and then autofill to P1.

2. To help differentiate the original data inside columns C:F from the Steps, it is a good idea to format the text words in G1:P1 such that they are bold and red (or another non-black color).

3. In cells G2:G11, enter numbers 11–20. These numbers represent the current clusters each data point is in. Note, since we have ten data points, any integer greater than ten can mark that this data point hasn't been clustered. This is the initial step, that is, Step-1.

At this moment, our worksheet should look like Figure 8-10.

	A	B	C	D	E	F	G	H	I	J	K	L	M	N	O	P
1	Fixed	Country	Murder	Assault	UrbanPop	Rape	Step-1	Step-2	Step-3	Step-4	Step-5	Step-6	Step-7	Step-8	Step-9	Step-10
2	1	L1	12	193	51	21	11									
3	2	L2	9	222	39	42	12									
4	3	L3	6	226	72	30	13									
5	4	L4	8	177	50	17	14									
6	5	L5	7	235	86	39	15									
7	6	L6	5	170	78	37	16									
8	7	L7	0	35	75	7	17									
9	8	L8	4	198	70	15	18									
10	9	L9	13	259	74	30	19									
11	10	L10	15	136	52	24	20									
12																

Figure 8-10. *The initial step of hierarchical clustering*

4. In cells B13:K13, enter integers 1–10, respectively. These numbers are used to help fetch a record from the array C2:F11.

5. In cells A14:A23, enter integers 1–10, respectively, too. We are going to construct the proximity matrix in cells B14:K23.

6. In cell B14, enter the formula

    ```
    =IFERROR(IF($A14=B$13,1000000, SQRT(SUMXMY2($C2:$F2,INDEX
    ($C$2:$F$11,B$13,0)))),1000000)
    ```

    In this formula, INDEX($C$2:$F$11,B$13,0) picks a record referenced by B13. In this particular case, it is the first data point (C2:F2).

    SQRT(SUMXMY2($C2:$F2,INDEX($C$2:$F$11,B$13,0))) computes the distance between C2:F2 and C2:F2.

Inside the IF function, the integer 1000000 is used as the maximum distance. If $A14=B$13, we know that it is computing the distance between the same data point (in this particular case, it is between C2:F2 and C2:F2). Though such a distance should be 0, but we would like to assign it the maximum distance for later use. Else the calculation result of the SQRT function will be returned.

The IFERROR function is used to make sure whenever there is an error, the maximum distance is assigned. This is important and necessary!

7. Autofill from B14 to K23. Now the proximity matrix has been constructed.

8. Enter the formula =MATCH(MIN($B$14:$K$23),B14:K14,0) in cell L14. Autofill from L14 to L23. The proximity matrix should look like Figure 8-11 and it tells that data points 4 and 1 have the shortest distance between.

13		1	2	3	4	5	6	7	8	9	10	
14	1	1000000	37.88139	40.58325	17	57.77543	39.53479	160.8726	22.04541	70.47695	57.16642	4
15	2	37.88139	1000000	35.4683	52.64979	48.89785	65.31462	193.8324	47.86439	52.47857	89.02247	#N/A
16	3	40.58325	35.4683	1000000	55.29919	18.9473	56.76266	192.4968	31.89044	33.79349	92.8278	#N/A
17	4	17	52.64979	55.29919	1000000	71.72866	35.24202	144.7515	29.3428	86.56789	42.22558	1
18	5	57.77543	48.89785	18.9473	71.72866	1000000	65.55151	202.9631	47.01064	28.93095	106.0472	#N/A
19	6	39.53479	65.31462	56.76266	35.24202	65.55151	1000000	138.416	36.51027	89.72179	45.83667	#N/A
20	7	160.8726	193.8324	192.4968	144.7515	202.9631	138.416	1000000	163.3218	225.5549	106.0377	#N/A
21	8	22.04541	47.86439	31.89044	29.3428	47.01064	36.51027	163.3218	1000000	63.58459	66.10598	#N/A
22	9	70.47695	52.47857	33.79349	86.56789	28.93095	89.72179	225.5549	63.58459	1000000	125.1119	#N/A
23	10	57.16642	89.02247	92.8278	42.22558	106.0472	45.83667	106.0377	66.10598	125.1119	1000000	#N/A

Figure 8-11. *The proximity matrix and the first cluster*

9. Step-2:

- Now we need to update the cluster information for the ten data points. Copy the numbers inside G2:G11 to H2:H11.

- Update the number inside H2 and H5 to 1. Here, 1 indicates this is the first cluster.

- Since countries L1 and L4 are inside cluster 1, we need to compute the centroid of cluster 1. In cell N14, enter the formula =AVERAGE(C2, C5) and then autofill to Q14. Be advised that a cluster is represented by its centroid in this exercise.

- Copy the content of N14:Q14 to N15:Q15 by values only. Note, it is very important to copy N14:Q14 and paste by values to N15:Q15. The purpose is to avoid accidental iterative calculation in next step.

- Copy N15:Q15 to C2:F2, and then clear the content in C5:F5. Our worksheet should look like Figure 8-12.

  Note, it is very important to clear the content in C5:F5 for country L4 because country L1 (C2:F2) is now representing cluster 1.

▲	A	B	C	D	E	F	G	H	I	J	K	L	M	N	O	P
1	Fixed	Country	Murder	Assault	UrbanPop	Rape	Step-1	Step-2	Step-3	Step-4	Step-5	Step-6	Step-7	Step-8	Step-9	Step-10
2	1	L1	10	185	50.5	19	11	1								
3	2	L2	9	222	39	42	12	12								
4	3	L3	6	226	72	30	13	13								
5	4	L4					14	1								
6	5	L5	7	235	86	39	15	15								
7	6	L6	5	170	78	37	16	16								
8	7	L7	0	35	75	7	17	17								
9	8	L8	4	198	70	15	18	18								
10	9	L9	13	259	74	30	19	19								
11	10	L10	15	136	52	24	20	20								
12																
13		1	2	3	4	5	6	7	8	9	10					
14	1	1000000	45.06939	47.75196	1000000	64.56973	36.47259	152.7883	24.5204	78.47452	49.53029	#N/A		10	185	50.5
15	2	45.06939	1000000	35.4683	1000000	48.89785	65.31462	193.8324	47.86439	52.47857	89.02247	#N/A		10	185	50.5
16	3	47.75196	35.4683	1000000	1000000	18.9473	56.76266	192.4968	31.89044	33.79349	92.8278		5			
17	4	1000000	1000000	1000000	1000000	1000000	1000000	1000000	1000000	1000000	1000000	#N/A				
18	5	64.56973	48.89785	18.9473	1000000	1000000	65.55151	202.9631	47.01064	28.93095	106.0472		3			
19	6	36.47259	65.31462	56.76266	1000000	65.55151	1000000	138.416	36.51027	89.72179	45.83667	#N/A				
20	7	152.7883	193.8324	192.4968	1000000	202.9631	138.416	1000000	163.3218	225.5549	106.0377	#N/A				
21	8	24.5204	47.86439	31.89044	1000000	47.01064	36.51027	163.3218	1000000	63.58459	66.10598	#N/A				
22	9	78.47452	52.47857	33.79349	1000000	28.93095	89.72179	225.5549	63.58459	1000000	125.1119	#N/A				
23	10	49.53029	89.02247	92.8278	1000000	106.0472	45.83667	106.0377	66.10598	125.1119	1000000	#N/A				

Figure 8-12. *After step-2, the cell array L14:L23 automatically shows that countries L5 and L3 are in a cluster together*

10. Step-3:

- Figure 8-12 shows that countries L3 and L5 are in a cluster. Since the current cluster values of L3 and L5 are 13 and 15, both greater than 10, we know that L3 and L5 haven't be clustered. Thus, they should be in cluster 2 together.

- Copy H2:H11 to I2:I11.

- Update cells I4 and I6 to 2.

- Time to compute the centroid of countries L3 and L5. In cell N14, enter =AVERAGE(C4,C6) and then autofill from N14 to Q14.

- Copy N14:Q14 and paste them by values into N15:Q15.

- Copy and paste N15:Q15 into C4:F4. We are using C4:F4 (country L3) to represent the cluster 2.

- Clear the content in C6:F6.

  Our worksheet should look like Figure 8-13.

	A	B	C	D	E	F	G	H	I	J	K	L	M	N	O	P	Q
1	Fixed	Country	Murder	Assault	UrbanPop	Rape	Step-1	Step-2	Step-3	Step-4	Step-5	Step-6	Step-7	Step-8	Step-9	Step-10	
2	1	L1	10	185	50.5	19	11	1	1								
3	2	L2	9	222	39	42	12	12	12								
4	3	L3	6.5	230.5	79	34.5	13	13	2								
5	4	L4					14	1	1								
6	5	L5					15	15	2								
7	6	L6	5	170	78	37	16	16	16								
8	7	L7	0	35	75	7	17	17	17								
9	8	L8	4	198	70	15	18	18	18								
10	9	L9	13	259	74	30	19	19	19								
11	10	L10	15	136	52	24	20	20	20								
12																	
13		1	2	3	4	5	6	7	8	9	10						
14	1	1000000	45.06939	55.99107	1000000	1000000	36.47259	152.7883	24.5204	78.47452	49.53029	8		6.5	230.5	79	34.5
15	2	45.06939	1000000	41.65033	1000000	1000000	65.31462	193.8324	47.86439	52.47857	89.02247	#N/A		6.5	230.5	79	34.5
16	3	55.99107	41.65033	1000000	1000000	1000000	60.57846	197.5721	39.03524	29.99583	99.20559	#N/A					
17	4	1000000	1000000	1000000	1000000	1000000	1000000	1000000	1000000	1000000	1000000	#N/A					
18	5	1000000	1000000	1000000	1000000	1000000	1000000	1000000	1000000	1000000	1000000	#N/A					
19	6	36.47259	65.31462	60.57846	1000000	1000000	1000000	138.416	36.51027	89.72179	45.83667	#N/A					
20	7	152.7883	193.8324	197.5721	1000000	1000000	138.416	1000000	163.3218	225.5549	106.0377	#N/A					
21	8	24.5204	47.86439	39.03524	1000000	1000000	36.51027	163.3218	1000000	63.58459	66.10598	1					
22	9	78.47452	52.47857	29.99583	1000000	1000000	89.72179	225.5549	63.58459	1000000	125.1119	#N/A					
23	10	49.53029	89.02247	99.20559	1000000	1000000	45.83667	106.0377	66.10598	125.1119	1000000	#N/A					

Figure 8-13. *Countries L3 and L5 are in cluster 2, Country L8 is in the same cluster as L1*

11. Step-4:

- Figure 8-13 also automatically shows that country L8 is in the same cluster as L1. As I2:I11 of Step-3 reveals that L8 has not been clustered (its current cluster value is 18) but L1 has, L8 should now be assigned to cluster 1 with L1.

- Copy I2:I11 to J2:J11; update J9 to 1.

- Recompute the centroid of cluster 1. Enter =SUM(C2*2,C9)/3 in cell N14; autofill from N14 to Q14. Note, because cluster 1 already has two data points and now includes country L8, we need to double C2 first.

- Copy N14:Q14 and paste by values to N15:Q15, then copy N15:Q15 to C2:F2.

- Clear the content inside C9:F9. Our worksheet should disclose that countries L3 and L9 are in the same cluster. Refer to Figure 8-14.

	A	B	C	D	E	F	G	H	I	J	K	L
1	Fixed	Country	Murder	Assault	UrbanPop	Rape	Step-1	Step-2	Step-3	Step-4	Step-5	Step-6
2	1	L1	8	189.3333	57	17.66667	11	1	1	1		
3	2	L2	9	222	39	42	12	12	12	12		
4	3	L3	6.5	230.5	79	34.5	13	13	2	2		
5	4	L4					14	1	1	1		
6	5	L5					15	15	2	2		
7	6	L6	5	170	78	37	16	16	16	16		
8	7	L7	0	35	75	7	17	17	17	17		
9	8	L8					18	18	18	1		
10	9	L9	13	259	74	30	19	19	19	19		
11	10	L10	15	136	52	24	20	20	20	20		
12												
13		1	2	3	4	5	6	7	8	9	10	
14	1	1000000	44.54461	49.64177	1000000	1000000	34.60572	155.9505	1000000	72.93528	54.39261	#N/A
15	2	44.54461	1000000	41.65033	1000000	1000000	65.31462	193.8324	1000000	52.47857	89.02247	#N/A
16	3	49.64177	41.65033	1000000	1000000	1000000	60.57846	197.5721	1000000	29.99583	99.20559	➡ 9
17	4	1000000	1000000	1000000	1000000	1000000	1000000	1000000	1000000	1000000	1000000	#N/A
18	5	1000000	1000000	1000000	1000000	1000000	1000000	1000000	1000000	1000000	1000000	#N/A
19	6	34.60572	65.31462	60.57846	1000000	1000000	1000000	138.416	1000000	89.72179	45.83667	#N/A
20	7	155.9505	193.8324	197.5721	1000000	1000000	138.416	1000000	1000000	225.5549	106.0377	#N/A
21	8	1000000	1000000	1000000	1000000	1000000	1000000	1000000	1000000	1000000	1000000	#N/A
22	9	72.93528	52.47857	29.99583	1000000	1000000	89.72179	225.5549	1000000	1000000	125.1119	➡ 3
23	10	54.39261	89.02247	99.20559	1000000	1000000	45.83667	106.0377	1000000	125.1119	1000000	#N/A

Figure 8-14. *After Step-4, countries L3 and L9 are in the same cluster*

12. Step-5:

- Certainly, country L9 should be in cluster 2 with L3. Copy J2:J11 to K2:K11, and update K10 to 2.

- To recompute the centroid of cluster 2, enter =SUM(C4*2, C10)/3 in N14; autofill from N14 to Q14.

- Copy N14:Q14 and paste by values into N15:Q15.

- Copy and paste N15:Q15 to C4:F4, then clear the content of C10:F10. The proximity matrix will reveal that country L6 is in the same cluster with L1.

13. Step-6:

- Copy K2:K11 to L2:L11; update cell L7 (note, it is the cell L7 not the Country L7!) to 1 to mark that country L6 is in cluster 1.

- To recompute the centroid for cluster 1, enter =SUM(C2*3, C7)/4 in N14; autofill from N14 to Q14.

- Copy/paste from N14:Q14 by values to N15:Q15, then copy/paste N15:Q15 to C2:F2.

- Clear the content inside C7:F7.

    Our worksheet should look like Figure 8-15.

▲	A	B	C	D	E	F	G	H	I	J	K	L
1	Fixed	Country	Murder	Assault	UrbanPop	Rape	Step-1	Step-2	Step-3	Step-4	Step-5	Step-6
2	1	L1	7.25	184.5	62.25	22.5	11	1	1	1	1	1
3	2	L2	9	222	39	42	12	12	12	12	12	12
4	3	L3	8.666667	240	77.33333	33	13	13	2	2	2	2
5	4	L4					14	1	1	1	1	1
6	5	L5					15	15	2	2	2	2
7	6	L6					16	16	16	16	16	1
8	7	L7	0	35	75	7	17	17	17	17	17	17
9	8	L8					18	18	18	1	1	1
10	9	L9					19	19	19	19	2	2
11	10	L10	15	136	52	24	20	20	20	20	20	20
12												
13			1	2	3	4	5	6	7	8	9	10
14	1	1000000	48.27137	58.48088	1000000	1000000	1000000	151.0153	1000000	1000000	50.19587	#N/A
15	2	48.27137	1000000	43.29614	1000000	1000000	1000000	193.8324	1000000	1000000	89.02247	3
16	3	58.48088	43.29614	1000000	1000000	1000000	1000000	206.837	1000000	1000000	107.6052	2
17	4	1000000	1000000	1000000	1000000	1000000	1000000	1000000	1000000	1000000	1000000	#N/A
18	5	1000000	1000000	1000000	1000000	1000000	1000000	1000000	1000000	1000000	1000000	#N/A
19	6	1000000	1000000	1000000	1000000	1000000	1000000	1000000	1000000	1000000	1000000	#N/A
20	7	151.0153	193.8324	206.837	1000000	1000000	1000000	1000000	1000000	1000000	106.0377	#N/A
21	8	1000000	1000000	1000000	1000000	1000000	1000000	1000000	1000000	1000000	1000000	#N/A
22	9	1000000	1000000	1000000	1000000	1000000	1000000	1000000	1000000	1000000	1000000	#N/A
23	10	50.19587	89.02247	107.6052	1000000	1000000	1000000	106.0377	1000000	1000000	1000000	#N/A

Figure 8-15. *After Step-6, countries L2 and L3 are in the same cluster*

14. Step-7:

- Copy the cell array L2:L11 to M2:M11.

- Since country L2 is in the same cluster with L3, it must be in cluster 2. Update M3 to 2.

- To recompute the centroid of cluster 2, enter =SUM(C4*3,C3)/4 in N14; autofill from N14 to Q14.

- Again, we need to copy N14:Q14 and paste by values into N15:Q15, then copy N15:Q15 into C4:F4.

- Clear the content of C3:F3.

15. Step-8:

- At this point, the proximity matrix reveals that countries L1 and L10 are in the same cluster. Let's copy M2:M11 to N2:N11 and update N11 to 1.

- In N14, enter the formula =SUM(C2*4,C11)/5 and autofill from N14 to Q14.

- Copy/paste N14:Q14 by values to N15:Q15, then copy N15:Q15 to C2:F2.

- After clearing the content in C11:F11, the proximity matrix shows that countries L3 and L1 are in the same cluster. This is demonstrated in Figure 8-16.

	A	B	C	D	E	F	G	H	I	J	K	L	M	N
1	Fixed	Country	Murder	Assault	UrbanPop	Rape	Step-1	Step-2	Step-3	Step-4	Step-5	Step-6	Step-7	Step-8
2	1	L1	8.8	174.8	60.2	22.8	11	1	1	1	1	1	1	1
3	2	L2					12	12	12	12	12	12	2	2
4	3	L3	8.75	235.5	67.75	35.25	13	13	2	2	2	2	2	2
5	4	L4					14	1	1	1	1	1	1	1
6	5	L5					15	15	2	2	2	2	2	2
7	6	L6					16	16	16	16	16	1	1	1
8	7	L7	0	35	75	7	17	17	17	17	17	17	17	17
9	8	L8					18	18	18	1	1	1	1	1
10	9	L9					19	19	19	19	2	2	2	2
11	10	L10					20	20	20	20	20	20	20	1
12														
13		1	2	3	4	5	6	7	8	9	10			
14	1	1000000	1000000	62.42193	1000000	1000000	1000000	141.7398	1000000	1000000	1000000	➝ 3		7.04
15	2	1000000	1000000	1000000	1000000	1000000	1000000	1000000	1000000	1000000	1000000	#N/A		8.8
16	3	62.42193	1000000	1000000	1000000	1000000	1000000	202.799	1000000	1000000	1000000	➝ 1		
17	4	1000000	1000000	1000000	1000000	1000000	1000000	1000000	1000000	1000000	1000000	#N/A		
18	5	1000000	1000000	1000000	1000000	1000000	1000000	1000000	1000000	1000000	1000000	#N/A		
19	6	1000000	1000000	1000000	1000000	1000000	1000000	1000000	1000000	1000000	1000000	#N/A		
20	7	141.7398	1000000	202.799	1000000	1000000	1000000	1000000	1000000	1000000	1000000	#N/A		
21	8	1000000	1000000	1000000	1000000	1000000	1000000	1000000	1000000	1000000	1000000	#N/A		
22	9	1000000	1000000	1000000	1000000	1000000	1000000	1000000	1000000	1000000	1000000	#N/A		
23	10	1000000	1000000	1000000	1000000	1000000	1000000	1000000	1000000	1000000	1000000	#N/A		

Figure 8-16. *After Step-8, countries L1 and L3 are inside the same cluster*

16. Step-9:

- Since country L1 is in cluster 1 and L3 is in cluster 2, both have been processed. This indicates that we need to merge cluster 1 and cluster 2 now. Thus, copy N2:N11 to O2:O11. Pay attention that there are five 1s and four 2s in O2:O11.

- Do not update O2:O11 at this moment until we have finished recomputing the centroid of the merged clusters. Enter =SUM(C2*5, C4*4)/9 in N14 and autofill from N14 to Q14.

- Copy N14:Q14 and paste by values into N15:Q15.

- Now update all the 2s inside O2:O11 to 1.

- Make sure that the values of N15:Q15 are copied to C2:F2, then clear the content of C4:F4.

17. Step-10:

- At this moment, the proximity matrix should tell that L7 is in the same cluster with L1.

- Thus, we can fill in 1s in P2:P11.

137

Our worksheet should look the same as Figure 8-17.

	A	B	C	D	E	F	G	H	I	J	K	L	M	N	O	P
1	Fixed	Country	Murder	Assault	UrbanPop	Rape	Step-1	Step-2	Step-3	Step-4	Step-5	Step-6	Step-7	Step-8	Step-9	Step-10
2	1	L1	8.777778	201.7778	63.55556	28.33333	11	1	1	1	1	1	1	1	1	1
3	2	L2					12	12	12	12	12	12	2	2	1	1
4	3	L3					13	13	2	2	2	2	2	2	1	1
5	4	L4					14	1	1	1	1	1	1	1	1	1
6	5	L5					15	15	2	2	2	2	2	2	1	1
7	6	L6					16	16	16	16	16	1	1	1	1	1
8	7	L7	0	35	75	7	17	17	17	17	17	17	17	17	17	1
9	8	L8					18	18	18	1	1	1	1	1	1	1
10	9	L9					19	19	19	19	2	2	2	2	1	1
11	10	L10					20	20	20	20	20	20	20	1	1	1
12																
13		1	2	3	4	5	6	7	8	9	10					
14	1	1000000	1000000	1000000	1000000	1000000	1000000	168.7541	1000000	1000000	1000000	7		4.8765	112.1	35.309
15	2	1000000	1000000	1000000	1000000	1000000	1000000	1000000	1000000	1000000	1000000	#N/A		8.777778	201.7778	63.55556
16	3	1000000	1000000	1000000	1000000	1000000	1000000	1000000	1000000	1000000	1000000	#N/A				
17	4	1000000	1000000	1000000	1000000	1000000	1000000	1000000	1000000	1000000	1000000	#N/A				
18	5	1000000	1000000	1000000	1000000	1000000	1000000	1000000	1000000	1000000	1000000	#N/A				
19	6	1000000	1000000	1000000	1000000	1000000	1000000	1000000	1000000	1000000	1000000	#N/A				
20	7	168.7541	1000000	1000000	1000000	1000000	1000000	1000000	1000000	1000000	1000000	1				
21	8	1000000	1000000	1000000	1000000	1000000	1000000	1000000	1000000	1000000	1000000	#N/A				
22	9	1000000	1000000	1000000	1000000	1000000	1000000	1000000	1000000	1000000	1000000	#N/A				
23	10	1000000	1000000	1000000	1000000	1000000	1000000	1000000	1000000	1000000	1000000	#N/A				

Figure 8-17. *All data points are clustered into cluster 1 at Step-10*

18. Our task stays unfinished. We need to organize the ten data points
    so that we can have a better idea how to draw the dendrogram. To
    do so, we need to sort the table B1:P11.

    Select B1:P11 ➤ click Data ➤ Sort ➤ at "Sort by", select Step-2 and then click
    OK. This is explained in Figure 8-18. Note, "My data has headers" is checked.

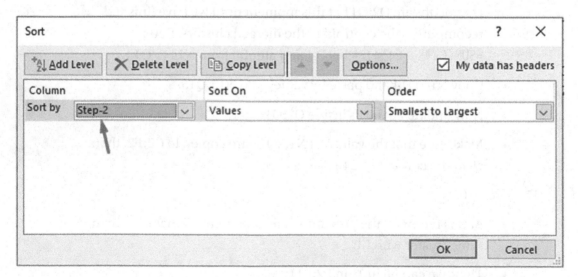

Figure 8-18. *Sort B1:P11 by Step-2. Note, "My data has headers" is checked*

19. At "Sort by", change Step-2 to Step-3 to sort B1:P11 by Step-3. Don't forget to click OK to complete the sorting process.

20. Continue sorting by Step-4, ..., and Step-9, respectively, and in order. Part of our worksheet should look like Figure 8-19.

▲	A	B	C	D	E	F	G	H	I	J	K	L	M	N	O	P
1	Fixed	Country	Murder	Assault	UrbanPop	Rape	Step-1	Step-2	Step-3	Step-4	Step-5	Step-6	Step-7	Step-8	Step-9	Step-10
2	1	L1	8.777778	201.7778	63.55556	28.33333	11	1	1	1	1	1	1	1	1	1
3	2	L4					14	1	1	1	1	1	1	1	1	1
4	3	L8					18	18	18	1	1	1	1	1	1	1
5	4	L6					16	16	16	16	16	1	1	1	1	1
6	5	L10					20	20	20	20	20	20	20	1	1	1
7	6	L3					13	13	2	2	2	2	2	2	1	1
8	7	L5					15	15	2	2	2	2	2	2	1	1
9	8	L9					19	19	19	19	2	2	2	2	1	1
10	9	L2					12	12	12	12	12	12	2	2	1	1
11	10	L7	0	35	75	7	17	17	17	17	17	17	17	17	17	1

Figure 8-19. *After sorting, the data is ready for drawing a dendrogram*

21. After sorting, the data is ready for drawing a dendrogram. Take a close look at the cell array B1:B11, which shows that when we draw our dendrogram, we should arrange our data points (countries) from left to right in this order: L1, L4, L8, L6, L10, L3, L5, L9, L2, L7.

Data inside G1:P11 instruct us the following:

- (L1, L4) is the first level of cluster.

- (L3, L5) is the second level of cluster.

- ((L1, L4), L8) is the third level of cluster.

- ((L3, L5), L9) is the fourth level of cluster.

- (((L1, L4), L8), L6) is the fifth level of cluster, and so on.

The current result without the dendrogram can be found in Chapter8-2b.xlsx. Please try to apply what you have learned regarding drawing a dendrogram to draw a dendrogram inside Chapter8-2b.xlsx.

Note, I completed the hierarchical clustering process inside the same worksheet. Should you want to keep the intermediate steps intact, you can make a copy of the worksheet at each of the intermediate steps, similar to what we did in k-means clustering. In fact, keeping intermediate steps intact can help avoid errors.

Chapter8-2e.xlsx repeats the preceding process by copying worksheets to keep intermediate steps intact. Please take a look if you are interested.

22. To draw the dendrogram, it is a good idea to delineate it on a piece of paper with a pen first (or on an iPad-like device). This will greatly help us specify the X and Y values for the upcoming scatter plot. I won't go over in detail how the X and Y values are generated, but they are available in Chapter8-2c.xlsx which you should open now.

23. In Chapter8-2c.xlsx, the X and Y values for our dendrogram is presented in cell array R1:S54. Please select them and draw a "X Y Scatter with Straight Lines". We shall get a chart like Figure 8-20.

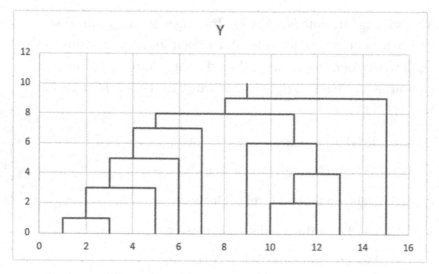

Figure 8-20. *The X Y Scatter with Straight Lines for our dendrogram*

24. Please revise the title to "Dendrogram".

25. Click on the x-axis to delete it.

26. Click "Add Chart Elements" ➤ Axis Titles.

27. Revise the y-axis title to "Distance not in proportion". As indicated by this y-axis title, the clade heights in our dendrogram do not reflect the proportional relationship of the cluster distances.

28. Revise the x-axis title to "L1 L4 L8 L6 L10 L3 L5 L9 L2 L7". After that, add or remove spaces between the words so that they align well with your chart.

This approach is different from applying data label. Certainly, you can try to add data labels instead. Figure 8-21 shows the dendrogram I created by adjusting x-axis title content.

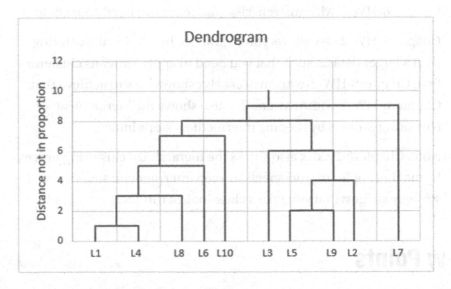

Figure 8-21. *One final dendrogram. Observe that the distances between the text words inside the x-axis title are explicitly enlarged*

The final result can be found in Chapter8-2d.xlsx which also draws a dendrogram by applying data labels.

Reinforcement Exercises

You may have realized that accomplishing hierarchical clustering and dendrogram via Excel is error prone because of the necessary manual work. This is especially true when the number of data points increases. As I mentioned before, one way to mitigate errors is to keep intermediate steps intact by making copies of the working worksheet, similar to what we did in k-means clustering.

Can we conduct hierarchical clustering on hundreds of data points through Excel? Certainly we can. But we must borrow a hand from k-means clustering first. The idea is to apply k-means clustering (e.g., k=7) on the larger dataset first, then treat each cluster as one data point to start hierarchical clustering.

- The file Chapter8-HW-1.xlsx provides an opportunity for you to practice hierarchical clustering and dendrogram drawing. Refer to Chapter8-HW-1-withAnswers.xlsx whenever you need some help.

- Chapter8-HW-2.xlsx allows you to practice hierarchical clustering with a larger dataset such that you need to apply k-means clustering first. Chapter8-HW-2-withAnswers.xlsx shows how to achieve that. Chapter8-HW-2-withAnswers.xlsx also shows the hierarchical clustering process by keeping intermediate steps intact.

- Note, Chapter8-2e.xlsx also shows the hierarchical clustering process by making copies of worksheets to keep intermediate steps intact. I strongly suggest that you take a close look at this file.

Review Points

For Chapter 8, the following knowledge, skills, and functions are worth reviewing:

1. Hierarchical clustering and dendrogram

2. The algorithm for agglomerative clustering

3. Proximity matrix

4. The Excel built-in features SORT and Paste by Values

5. Scatter plot chart and data labels

6. Functions IF, MIN, IFERROR, INDEX, MATCH, SQRT, and SUMXMY2

7. Copy worksheet to keep intermediate steps intact

CHAPTER 9

Naive Bayes Classification

The data mining methods we have been learning all favor numerical data: linear regression, LDA, k-means clustering, logistic regression, K-NN, and hierarchical clustering. Naïve Bayes favors categorical data, however. Because of its simplicity, Naïve Bayes data mining method is much more efficient compared to many other data mining methods, while its performance can still match most other data mining methods.

Please download the sample Excel files from `https://github.com/hhohho/Learn-Data-Mining-through-Excel-2` for this chapter's exercises.

General Understanding

Data mining methods or machine learning models are generally considered probability methods/models. The probability computation starts from the training dataset. A fundamental principal is that if a specific state of the attribute or the target S_k has the probability P_k to appear in a given condition, then it is assumed that under the same condition in the scoring dataset, the probability for S_k to appear is still P_k. When you read this chapter and other chapters such as Association Analysis and Decision Trees, you will find the implicit application of this principal, though I may or may not explicitly mention it.

Naive Bayes (Naïve Bayes) is one of the most used classification algorithms. Its mathematical background is the conditional probability which asks this question: what is the probability for event y to happen given x happens? Here, the condition is x (x happens or x is true), and the conditional probability is expressed as $P(y|x)$.

The conditional probability $P(y|x)$ is computed as

$$P(y|x) = \frac{P(y \cap x)}{P(x)}, \qquad P(x) \neq 0 \tag{9-1}$$

© Hong Zhou 2023
H. Zhou, *Learn Data Mining Through Excel*, https://doi.org/10.1007/978-1-4842-9771-1_9

where $P(x)$ represents the probability for x to be true and $P(y \cap x)$ is the probability for both x and y to be true.

Similarly, the conditional probability $P(x|y)$ is computed as

$$P(x|y) = \frac{P(x \cap y)}{P(y)}, \qquad P(y) \neq 0 \qquad (9\text{-}2)$$

Because $P(y \cap x) = P(x \cap y)$, we then have $P(y|x)P(x) = P(x|y)P(y)$. Thus, we can derive the simplest Bayes' theorem as stated in Equation (9-3):

$$P(y|x) = \frac{P(x|y)P(y)}{P(x)}, \qquad P(x) \neq 0 \qquad (9\text{-}3)$$

Note, the Bayes' theorem converts the computation of $P(y|x)$ into $P(x|y)$.

Equation (9-3) implicitly assumes that y is the class in study for the dependent variable y and x is the state in study for the independent attribute x. This may be a little bit confusing since y must have more than one class and x must have more than one state. Don't worry, please read on and you will develop a clear understanding of the Bayes' theorem.

Assume that y has two classes y_1 and y_2, let's rewrite Equation (9-3) for a given y_1 under a specific condition x_a (assume x has different states/conditions and x_a is one of them), we then have

$$P(y_1|x_a) = \frac{P(x_a|y_1)P(y_1)}{P(x_a)}, \qquad P(x_a) \neq 0 \qquad (9\text{-}4)$$

Repeat the preceding action for y_2; we have

$$P(y_2|x_a) = \frac{P(x_a|y_2)P(y_2)}{P(x_a)}, \qquad P(x_a) \neq 0 \qquad (9\text{-}5)$$

Because y has only two classes, under the condition x_a, either y_1 or y_2 must happen, that is, $P(y_1|x_a) + P(y_2|x_a)$ must equal 1. Therefore,

$$\frac{P(x_a|y_1)P(y_1)}{P(x_a)} + \frac{P(x_a|y_2)P(y_2)}{P(x_a)} = 1$$

that is, $P(x_a) = P(x_a|y_1)P(y_1) + P(x_a|y_2)P(y_2)$. Thus, we can finally reach

$$P(y_1|x_a) = \frac{P(x_a|y_1)P(y_1)}{P(x_a|y_1)P(y_1) + P(x_a|y_2)P(y_2)} \tag{9-6}$$

The beauty of Bayes' theorem is that it converts the computation of $P(y_1|x_a)$ into $P(x_a|y_1)$ and $P(x_a|y_2)$! Thinking about this in data mining. Given the training dataset where y is known, we can easily compute $P(x_a|y_1)$, $P(x_a|y_2)$, $P(y_1)$, and $P(y_2)$. Now for the scoring dataset where y is unknown but x is given. If x is x_a, we can then make use of Equation (9-6) to predict the occurrence probability of y_1 under the condition of x_a.

Equation (9-6) assumes that there is only one independent attribute x. In most cases, there are a number of independent attributes $(x_1, x_2, \ldots, x_n)$ to classify y and y can have more than two classes. Assume that y has m distinct classes and given the condition $(x_1, x_2, \ldots, x_n)$, let

$$P'(y_k|x) = P(x_1|y_k)P(x_2|y_k)\cdots P(x_n|y_k)P(y_k) \tag{9-7}$$

Following what we have gone through earlier, it is not difficult to deduce that the final $P(y_k|x)$, the probability of the kth y class under the condition of $(x_1, x_2, \ldots, x_n)$, can be computed as

$$P(y_k|x) = \frac{P'(y_k|x)}{\sum_i^m P'(y_i|x)} \tag{9-8}$$

Certainly, $\sum_i^m P(y_i|x) = 1$.

We are going to implement the preceding concepts using the example presented in Table 9-1.

Data in Table 9-1 simulate an image analysis case in which the occurrence of a certain cancer is determined based on features (attributes) of the images taken. Note, data in the table are faked and do not represent a valid image or cancer analysis method.

Table 9-1. *A faked sample data for Naive Bayes analysis*

No.	Size	Number	Thickness	Lung Cancer
1	Big	Low	Deep	Yes
2	Big	Over	Deep	Yes
3	Normal	Over	Shallow	No
4	Big	Low	Shallow	No
5	Big	Over	Deep	Yes
6	Normal	Over	Deep	Yes
7	Big	Low	Shallow	No
8	Big	Over	Shallow	Yes
9	Big	Over	Deep	Yes
10	Normal	Low	Deep	No
Scoring	Big	Over	Deep	???

There are three attributes: Size, Number, and Thickness (think them as x_1, x_2, and x_3). Each attribute has two distinct values. The target variable is Lung Cancer, which has two distinct classes: Yes and No. In Table 9-1, there are ten training data records, and the last record is the scoring record for which we need to predict the class based on the three attribute values.

Based on Equations (9-7) and (9-8), the probability of being "yes" for the scoring record is computed by the following equation:

$$P\big(\text{yes}|\langle \text{big}, \text{over}, \text{deep}\rangle\big) =$$

$$\frac{p(\text{big}|\text{yes})\times p(\text{over}|\text{yes})\times p(\text{deep}\,|\,\text{yes})\times p(\text{yes})}{p(\text{big}|\text{yes})\times p(\text{over}|\text{yes})\times p(\text{deep}\,|\,\text{yes})\times p(\text{yes})+p(\text{big}\,|\text{no})\times p(\text{over}\,|\text{no})\times p(\text{deep}|\text{no})\times p(\text{no})}$$

Certainly, we need to compute the following items:

1. p(yes) = 6/10, as there are 6 "yes" out of 10 records.

2. p(no) = 4/10, as there are 4 "no" out of 10 records.

3. p(big|yes) = 5/6. When the class is "yes", correspondingly there are 5 "big".

4. p(over|yes) = 5/6. When the class is "yes", correspondingly there are 5 "over".

5. p(deep|yes) = 5/6. When the class is "yes", correspondingly there are 5 "deep".

6. p(big|no) = 2/4. When the class is "no", correspondingly there are 2 "big".

7. p(over|no) = 1/4. When the class is "no", correspondingly there is 1 "over".

8. p(deep|no) = 1/4. When the class is "no", correspondingly there is 1 "deep".

Thus,

$$P\big(yes|\langle big, over, deep \rangle\big) =$$

$$\frac{(5/6)\times(5/6)\times(5/6)\times(6/10)}{(5/6)\times(5/6)\times(5/6)\times(6/10)+(2/4)\times(1/4)\times(1/4)\times(4/10)}$$

The final answer is 0.965. This says that the probability for the scoring record to be "yes" (lung cancer positive) is 0.965.

Certainly, the mathematical background of Naïve Bayes classification is relatively much simpler than that of logistic regression, LDA, and K-NN. This is an important reason why Naïve Bayes classification can have outstanding computation efficiency when the data size is large. In addition, though Naïve Bayes makes a "naïve" assumption that all the attributes are independent from each other, Naïve Bayes does have fair performance compared to many other classification data mining models.

Learn Naïve Bayes Through Excel

We are going to practice two examples through Excel in this chapter. The first example is the same as Table 9-1. The second example simulates a similar image analysis case. However, the data are more complicated, and the data size is much larger.

Exercise 1

The file Chapter9-1a.xlsx contains the same data as shown in Table 9-1. There are only ten training samples and one scoring sample. The reason to make the sample size small is for better illustration. Once we have gained enough experience through this simple example, we will work on another dataset which is more complicated.

In Chapter9-1a.xlsx, data are already set up in the worksheet for easy processing as shown in Figure 9-1. What we need to do is to complete necessary computations for Naïve Bayes analysis.

◢	A	B	C	D	E
1	Sample size		Lung Cancer		
2			yes	no	
3		count			
4		probability			
5	Size	big			
6		normal			
7	Number	over			
8		low			
9	Thickness	deep			
10		shallow			
11					
12	No.	Size	Number	Thickness	Lung Cancer
13	1	big	low	deep	yes
14	2	big	over	deep	yes
15	3	normal	over	shallow	no
16	4	big	low	shallow	no
17	5	big	over	deep	yes
18	6	normal	over	deep	yes
19	7	big	low	shallow	no
20	8	big	over	shallow	yes
21	9	big	over	deep	yes
22	10	normal	low	deep	no

Figure 9-1. *Naive Bayes data setup*

To apply Naïve Bayes method to predict the class of the scoring record, we need to build the model based on the training data first. Follow these instructions:

1. Enter 10 in cell B1, as there are only ten training records.

2. Enter the formula =COUNTIF($E$13:$E$22,C$2) in cell C3. This counts the number of class "yes" for the target variable Lung Cancer. Autofill from cell C3 to cell D3. Cell D3 stores the number of class "no".

3. Enter the formula =C$3/$B$1 in cell C4, and then autofill from cell C4 to cell D4. Cells C4 and D4 reference p(yes) and p(no).

4. Enter the following formula in cell C5:

   =COUNTIFS($E$13:$E$22,C$2,$B$13:$B$22,$B5)/C$3

5. Autofill from cell C5 to cell C6, and then autofill together to cells D5:D6. Cells C5, C6, D5, and D6 represent the probabilities p(big|yes), p(normal|yes), p(big|no), and p(normal|no), respectively. Recall p(big|yes) is the conditional probability: given "yes", the probability of Size = "big".

   At this moment, part of our worksheet looks like Figure 9-2.

◢	A	B	C	D
1	Sample size	10	Lung Cancer	
2			yes	no
3		count	6	4
4		probability	0.6	0.4
5	Size	big	0.8333333	0.5
6		normal	0.1666667	0.5
7	Number			
8				
9	Thickness			
10				

Figure 9-2. *Naive Bayes analysis is partially completed*

6. Enter the following formula in cell C7:

   =COUNTIFS($E$13:$E$22,C$2,$C$13:$C$22,$B7)/C$3

7. Autofill from cell C7 to cell C8, and then autofill together to cells D7:D8. Cells C7, C8, D7, and D8 reference the probabilities p(over|yes), p(low|yes), p(over|no), and p(low|no), respectively.

8. Enter the following formula in cell C9:

   =COUNTIFS($E$13:$E$22,C$2,$D$13:$D$22,$B9)/C$3

9. Autofill from cell C9 to cell C10, and then autofill together to cells D9:D10. Cells C9, C10, D9, and D10 reference the probabilities p(deep|yes), p(shallow|yes), p(deep|no), and p(shallow|no), respectively.

By now, our worksheet should look the same as Figure 9-3. If there are differences, examine the formulas inside the cells C3:D10.

◢	A	B	C	D	E
1	Sample size	10	Lung Cancer		
2			yes	no	
3		count	6	4	
4		probability	0.6	0.4	
5	Size	big	0.8333333	0.5	
6		normal	0.1666667	0.5	
7	Number	over	0.8333333	0.25	
8		low	0.1666667	0.75	
9	Thickness	deep	0.8333333	0.25	
10		shallow	0.1666667	0.75	
11					
12	No.	Hotspot-size	Number	Thickness	Lung Cancer

Figure 9-3. *Naïve Bayes data preparation*

Continue our Naïve Bayes data mining process by following these instructions:

10. Enter "p'(scoring|yes)", "p'(scoring|no)", and "P(yes|scoring)" in cells F22, G22, and H22, respectively.

11. Enter the following formula in cell F23:

```
=SUMIF($B$5:$B$10,$B23,C$5:C$10) * SUMIF($B$5:$B$10,$C23,C$5:C$10)
* SUMIF($B$5:$B$10,$D23,C$5:C$10)*C$4
```

The preceding formula computes the probability of the scoring record given the class is "yes". It implements Equation (9-7). Note that the function SUMIF has a different syntax from SUMIFS. For example, SUMIF ($B$5:$B$10,$B23,C$5:C$10) sums the cells in the array C5:C10 only if their matching cells in the array B5:B10 have the same value as cell B23. If we want to use the function SUMIFS for the same purpose, then the expression should be SUMIFS(C$5:C$10,$B$5:$B$10,$B23).

12. Autofill from cell F23 to cell G23. Cell G23 references the probability of the scoring record given the class is "no".

13. Enter in cell H23 the formula =F23/(F23+G23) which implements Equation (9-8).

14. Enter in cell E23 the formula =IF(H23>0.5,"yes","no").

This completes our first endeavor on the Naïve Bayes classification. Our result should look the same as what is shown in Figure 9-4.

The complete result is available in the file Chapter9-1b.xlsx.

⬚	A	B	C	D	E	F	G	H
12	No.	Size	Number	Thickness	Lung Cancer			
13	1	big	low	deep	yes			
14	2	big	over	deep	yes			
15	3	normal	over	shallow	no			
16	4	big	low	shallow	no			
17	5	big	over	deep	yes			
18	6	normal	over	deep	yes			
19	7	big	low	shallow	no			
20	8	big	over	shallow	yes			
21	9	big	over	deep	yes			
22	10	normal	low	deep	no	p'(scoring|yes)	p'(scoring|no)	P(yes|scoring)
23	Scoring	big	over	deep	yes	0.347222222	0.0125	0.965250965

Figure 9-4. *Naive Bayes classification completed*

Exercise 2

Hopefully by now you have developed some decent understanding about how to conduct a Naïve Bayes classification in Excel. Let's proceed with a more complicated example.

Open the file Chapter9-2a.xlsx which contains randomly generated data similar to Chapter9-1a.xlsx. In Chapter9-2a.xlsx, there are 1978 samples in the training dataset, and there are four distinct classes for the target variable "Lung Cancer". The four classes are negative, stage-1, stage-2, and stage-3. In addition, the three predictors all have more than two categorical values. Observe that Naïve Bayes method only works with categorical values. If our data values are not categorical but numerical, we must transform them into categorical values before applying Naïve Bayes classification method.

Take a close look at the data. Part of our worksheet looks like Figure 9-5. The training dataset ranges from row 34 to row 2011. The scoring dataset ranges from row 21 to row 30.

▲	A	B	C	D	E	F
1	Sample size	1978		Lung Cancer		
2			negative	stage-1	stage-2	stage-3
3		count				
4		probability				
5		ex-large				
6		large				
7	Size	median				
8		small				
9		none				
10		scale-1				
11		scale-2				
12	Number	scale-3				
13		scale-4				
14		scale-5				
15		light				
16	Thickness	grey				
17		dark				

Figure 9-5. *Setup of Exercise 2*

Follow these instructions to complete Naïve Bayes classification on the given dataset:

1. Enter the following formula in cell C3 and then autofill from cell C3 to cell F3:

    =COUNTIF($E$34:$E$2011,C2)

2. Enter the formula =C3/$B$1 in cell C4 and autofill from cell C4 to F4. Cells C4:F4 reference the probability of each class.

3. To compute p(ex-large|negative), enter the following formula in cell C5:

```
=COUNTIFS($B$34:$B$2011,$B5,$E$34:$E$2011,C$2)/C$3
```

4. Autofill from cell C5 to F5, then autofill together to cells C9:F9.

At this moment, examine if our worksheet looks similar to what is shown in Figure 9-6. Depending on the number formatting, the last couple digits of the decimal numbers in cells C4:F9 in our worksheet may be different from those in Figure 9-6. Make sure that our formulas are correct before continuing.

	A	B	C	D	E	F
1	Sample size	1978		Lung Cancer		
2			negative	stage-1	stage-2	stage-3
3		count	491	480	502	505
4		probability	0.248230536	0.24266936	0.253791709	0.25530839
5		ex-large	0.187372709	0.21666667	0.201195219	0.20594059
6		large	0.181262729	0.21875	0.187250996	0.22178218
7	Size	median	0.236252546	0.19583333	0.167330677	0.2019802
8		small	0.187372709	0.19583333	0.215139442	0.19207921
9		none	0.207739308	0.17291667	0.229083665	0.17821782
10		scale-1				
11		scale-2				
12	Number	scale-3				
13		scale-4				
14		scale-5				
15		light				
16	Thickness	grey				
17		dark				

Figure 9-6. *Examine our current data*

5. To compute p(scale-1|negative), enter the following formula in cell C10:

```
=COUNTIFS($C$34:$C$2011,$B10,$E$34:$E$2011,C$2)/C$3
```

6. Autofill from cell C10 to cell F10, and then autofill together to cells C14:F14.

7. To compute p(light|negative), enter the following formula in cell C15:

    ```
    =COUNTIFS($D$34:$D$2011,$B15,$E$34:$E$2011,C$2)/C$3
    ```

8. Autofill from cell C15 to cell F15, and then autofill together to cells C17:F17.

    Examine our worksheet again. It should look similar to Figure 9-7. By far, we have completed the computation for each individual conditional probability based on the training dataset.

▲	A	B	C	D	E	F
1	Sample size	1978	Lung Cancer			
2			negative	stage-1	stage-2	stage-3
3		count	491	480	502	505
4		probability	0.248230536	0.24266936	0.253791709	0.25530839
5		ex-large	0.187372709	0.21666667	0.201195219	0.20594059
6		large	0.181262729	0.21875	0.187250996	0.22178218
7	Size	median	0.236252546	0.19583333	0.167330677	0.2019802
8		small	0.187372709	0.19583333	0.215139442	0.19207921
9		none	0.207739308	0.17291667	0.229083665	0.17821782
10		scale-1	0.203665988	0.20625	0.211155378	0.21386139
11		scale-2	0.199592668	0.20416667	0.187250996	0.20792079
12	Number	scale-3	0.211812627	0.20625	0.203187251	0.1980198
13		scale-4	0.211812627	0.2	0.199203187	0.18613861
14		scale-5	0.17311609	0.18333333	0.199203187	0.19405941
15		light	0.356415479	0.36041667	0.338645418	0.32475248
16	Thickness	grey	0.364562118	0.2875	0.348605578	0.33069307
17		dark	0.279022403	0.35208333	0.312749004	0.34455446

Figure 9-7. *Examine our data and formula again*

Continue to work on the scoring dataset by following these instructions:

9. In cell F21, enter the following formula:

    ```
    =SUMIF($B$5:$B$17,$B21,C$5:C$17) * SUMIF($B$5:$B$17,
    $C21,C$5:C$17) * SUMIF($B$5:$B$17,$D21,C$5:C$17) * C$4
    ```

    Row 21 is the first scoring record. B21 is the Size attribute of the first scoring record. In this specific case, the value of B21 is "none". Cells C5:C17 store all the individual conditional probabilities with respect to class "negative".

    The expression `SUMIF($B$5:$B$17,$B21,C$5:C$17)` sums all the cells in array C5:C17 as long as their corresponding cells in array B5:B17 have the value "none". Indeed, this expression fetches the conditional probability p(none|negative) for scoring record 1 (attribute Size).

    `SUMIF($B$5:$B$17,$C21,C$5:C$17)` fetches the conditional probability p(scale-4|negative) for scoring record 1 (attribute Number).

    `SUMIF($B$5:$B$17,$D21,C$5:C$17)` fetches the conditional probability p(light|negative) for scoring data record 1 (attribute Thickness).

    The formula in cell F21 implements Equation (9-7) for scoring data record 1 when the class is "negative".

10. Autofill from cell F21 to cell I21.

    The formula in cell G21 implements Equation (9-7) for scoring data record 1 when the class is "stage-1".

    The formula in cell H21 implements Equation (9-7) for scoring data record 1 when the class is "stage-2".

    The formula in cell I21 implements Equation (9-7) for scoring data record 1 when the class is "stage-3".

11. In cell J21, enter the formula =MAX(F21:I21).

As there are more than two classes, the final categorical class is determined by the maximum among F21:I21. Note the final probability is not computed here, that is, Equation (9-8) is not implemented. If needed, we can compute the P(negative|record1) by the formula =F21/SUM(F21:I21). The probabilities of other classes for scoring record 1 can be computed in a similar way.

Part of our worksheet looks like Figure 9-8.

▲	A	B	C	D	E	F	G	H	I	J
19			Scoring Data							
20	SampleID	Size	Number	Thickness	Lung Cancer	P(t|n)	P(t|1)	P(t|2)	P(t|3)	P(max)
21	1	none	scale-4	light		0.00389298	0.003025	0.003922	0.00275	0.003922
22	2	large	scale-1	light						
23	3	small	scale-2	grey						
24	4	median	scale-5	dark						
25	5	none	scale-1	dark						
26	6	ex-large	scale-3	light						
27	7	none	scale-5	dark						
28	8	small	scale-1	grey						
29	9	median	scale-4	light						
30	10	small	scale-2	grey						

Figure 9-8. *Computing probabilities for scoring dataset*

12. Autofill cells F21:J21 together to cells F30:J30.

13. In cell E21, enter the following formula and then autofill from cell E21 to E30:

```
=INDEX($C$2:$F$2, MATCH(J21, F21:I21, 0)) &" ("&TEXT(J21/
SUM(F21:I21),"0.0000")&")"
```

The preceding formula is worth a detailed explanation. This formula finds the most likely corresponding class for scoring record 1 and displays the probability of this class.

a. Cells C2:F2 store the class names of the target variable Lung Cancer. Cells F21:I21 store the probabilities of scoring record 1 with respect to each class.

b. =INDEX($C$2:$F$2, MATCH(J21, F21:I21, 0)) finds the corresponding class of "Lung Cancer" based on the value inside cell J21: the maximum

157

of F21:I21. There must be a value inside F21:I21 matching cell J21. In this
specific case, it is H21. Function INDEX returns the value of the cell in
C2:F2 that aligns to H21 in F21:I21.

c. Operator & is used to concatenate multiple strings together.

d. J21/SUM(F21:I21) computes the probability of the most likely class of
scoring record 1 by following Equation (9-8).

e. TEXT(J21/SUM(F21:I21),"0.0000") formats the probability so that four
digits are displayed after the decimal point.

Our final result should look similar to Figure 9-9. Note as the training dataset is
randomly generated, the probabilities (column E) in Figure 9-9 are not dominant. They
are all close to 0.25.

19	Scoring Data				
20	SampleID	Size	Number	Thickness	Lung Cancer
21	1	none	scale-4	light	stage-2 (0.2886)
22	2	large	scale-1	light	stage-1 (0.2713)
23	3	small	scale-2	grey	stage-2 (0.2719)
24	4	median	scale-5	dark	stage-3 (0.2875)
25	5	none	scale-1	dark	stage-2 (0.2915)
26	6	ex-large	scale-3	light	stage-1 (0.2730)
27	7	none	scale-5	dark	stage-2 (0.3053)
28	8	small	scale-1	grey	stage-2 (0.2921)
29	9	median	scale-4	light	negative (0.3200)
30	10	small	scale-2	grey	stage-2 (0.2719)

Figure 9-9. *Naive Bayes classification is completed*

From Exercise 2, we can tell that Excel is very suitable for Naïve Bayes analysis. The
complete result of Exercise 2 can be found in Chapter9-2b.xlsx.

Note for the three attributes Size, Number, and Thickness, we entered three different
formulas at steps 3, 5, and 7. There is a way that we can complete the probability
computation for all three attributes by entering a formula in cell C5 and then autofill to
cell F17. We will experience such a technique in the next chapter.

The data presented in Chapter9-2a.xlsx are randomly generated; they are not
suitable for cross-validation. However, given the size of the training dataset, we can
certainly divide it into two parts to practice cross-validation.

Reinforcement Exercises

The data in Chapter9-HW.xlsx is downloaded from the famous UCI Machine Learning Repository at `https://archive.ics.uci.edu/`. This dataset provides a good opportunity for practicing Naïve Bayes data mining. Refer to Chapter9-HW-withAnswers whenever you need help.

Review Points

1. General understanding of Naïve Bayes. Learn how to deduce Bayes' theorem from conditional probability and try to repeat the deduction process by yourself.

2. COUNTIF, COUNTIFS, and SUMIF functions and operator &.

3. INDEX, MATCH, and TEXT functions.

CHAPTER 10

Decision Trees

Decision tree is probably the most intuitive data classification and prediction method. It is also used frequently. While most of the data mining methods we have learned are parametric, decision tree is a rule-based method. The most critical concept in understanding decision trees is entropy which will be explained soon. A tree is composed of nodes and the leaves are the bottom nodes. At each node except for the leaf nodes, a decision must be made to split the node into at least two branches. Figure 10-1 depicts a sample decision tree structure.

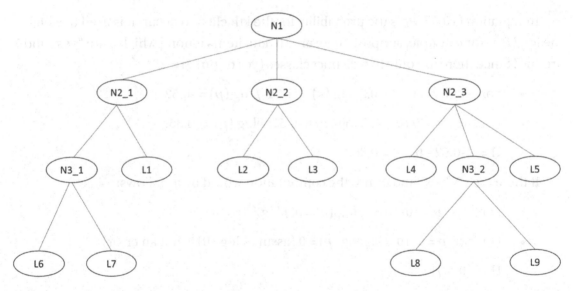

Figure 10-1. *A sample decision tree structure*

In Figure 10-1, all nodes start with letter "N", while all leaves start with letter "L". In a decision tree, a node can be split into two or more child nodes. However, a leaf can never be split again. The most critical action in a decision tree is to determine how to split a node. Though there is a general rule applied to every node, each node must be considered individually based on its inside data.

© Hong Zhou 2023
H. Zhou, *Learn Data Mining Through Excel*, https://doi.org/10.1007/978-1-4842-9771-1_10

Please download the sample Excel files from https://github.com/hhohho/Learn-Data-Mining-through-Excel-2 for this chapter's exercises.

General Understanding

How a node is split in a decision tree is based on the entropy computed with the data inside the node. Entropy represents how "pure" the data inside a tree node is. The higher the entropy, the less pure the data is. Let's learn how to calculate entropy so that we can develop a better understanding.

Assume that the target variable of a dataset has m distinct classes. The entropy of this dataset is usually computed based on Equation (10-1):

$$H = -\sum_{k=1}^{m} P_K \log_2\left(P_k\right) \qquad (10\text{-}1)$$

In Equation (10-1), P_K is the probability for the kth class to occur; it is used to weigh the $\log_2(P_k)$. For example, suppose there are 10 data items among which 3 are "yes" and 7 are "no". Since there are only two distinct classes (yes or no), m = 2.

- For "yes", $p = 3/10 = 0.3$, $\log_2(p) = -1.74$, $p\log_2(p) = -0.52$.

- For "no", $p = 7/10 = 0.7$, $\log_2(p) = -0.51$, $p\log_2(p) = -0.36$.

- H = -(-0.52 - 0.36) = 0.88.

If there are 10 "yes" and 0 "no", the computation would be as follows:

- For "yes", $p = 10/10 = 1$, $\log_2(p) = 0$, $p\log_2(p) = 0$.

- For "no", $p = 0/10 = 0$, $p\log_2(p) = 0$ (assume $\log_2(0)$ is not an error).

- H = -(0 + 0) = 0.

If there are 5 "yes" and 5 "no", the computation will give out H = 1.0.

It seems if the data are more skewed to a class, the entropy is lower. Decision tree method tends to pick the attribute that generates the least entropy to split a node. This is easy to understand. Suppose there are two paths before us, and we need to select one to reach our destination. If both paths offer us the same likelihood to arrive on time, we will find ourselves hesitating (H=1). If one way is definitely much more likely to allow us to arrive on time, our decision won't be difficult. Moreover, if there is only one way, our selection is definite (H = 0).

In Equation (10-1), base 2 logarithm is used. Base 2 logarithm is the most commonly used logarithm function in decision trees. When there are only two classes in a dataset, base 2 logarithm guarantees that the entropy is between 0 and 1 (inclusively). When there are more than two classes, base 2 logarithm cannot guarantee the entropy to be between 0 and 1. If we prefer the entropy to be between 0 and 1, then when there are N distinct classes, a logarithm function of base N must be used.

Let's use the well-known golf dataset (also known as the weather dataset) to explain how a decision tree is constructed.

Table 10-1. *The play golf dataset*

Temperature	Humidity	Windy	Outlook	Play
Hot	High	FALSE	overcast	Yes
Cool	Normal	TRUE	overcast	Yes
Mild	High	TRUE	overcast	Yes
Hot	Normal	FALSE	overcast	Yes
Mild	High	FALSE	rainy	Yes
Cool	Normal	FALSE	rainy	Yes
Cool	Normal	TRUE	rainy	No
Mild	Normal	FALSE	rainy	Yes
Mild	High	TRUE	rainy	No
Hot	High	FALSE	sunny	No
Hot	High	TRUE	sunny	No
Mild	High	FALSE	sunny	No
Cool	Normal	FALSE	sunny	Yes
Mild	Normal	TRUE	sunny	Yes

This dataset is very simple. It has only 14 samples and 4 attributes. Either playing golf or not depends on the 4 attributes. To build a decision tree based on this training dataset, the first job is to find an attribute to split the tree into multiple branches from the beginning. To do so, we need to compute the entropy for the target variable Play and entropies for the 4 attributes with respect to the target variable.

For the target variable Play, there are 9 Yes and 5 No; thus, by Equation (10-1)

```
H-play = -(9/14×log2(9/14)+5/14×log2(5/14)) = 0.94
```

For the attribute Outlook, there are 3 values (states): overcast, rainy, and sunny; each has 4, 5, and 5 data points. Considering overcast, the 4 overcast match to 4 Yes but 0 No. Thus,

```
H-outlook-overcast = -(4/4×log2(4/4)+0/4×log2(0/4)) = 0.0
```

Note, log2(0) = 0 here.

Similarly, the 5 rainy match to 3 Yes and 2 No, and the 5 sunny match to 2 Yes and 3 No. Thus,

```
H-outlook-rainy = -(3/5×log2(3/5)+2/5×log2(2/5)) = 0.97

H-outlook-sunny = -(2/5×log2(2/5)+3/5×log2(3/5)) = 0.97
```

The three sub-entropy values must be weighted before they are summed. The weights for overcast, rainy, and sunny are 4/14, 5/14, and 5/14, respectively. Thus,

```
H-outlook = 4/14×0.0+5/14×0.97+5/14×0.97 = 0.69
```

Continue this calculation for Temperature, Humidity, and Windy; we get

```
                    H-temperature = 0.91

                    H-humidity = 0.79

                    H-windy = 0.89
```

Here comes another concept: "information gain", which represents how much information can be gained by splitting the current dataset based on an attribute. It is defined as the difference between the entropy of the target variable and the entropy of one attribute. For example, the information gain for Outlook = 0.94 – 0.69 = 0.25.

Information gain can be biased to an attribute with more values. To reduce such a bias, "gain ratio" can be used. Gain ratio for an attribute is defined as

attribute gain ratio = attribute-information-gain / attribute-intrinsic-information

The intrinsic information for an attribute is calculated by Equation (10-2):

$$S = -\sum_{k=1}^{C} \frac{N_k}{N} \log_2 \left(\frac{N_k}{N} \right) \qquad (10\text{-}2)$$

In Equation (10-2), N represents the data size, N_k is the number of data points for a given attribute value, while C is the number of distinct values for an attribute.

Let's use attribute Outlook as the example again to implement Equation (10-2).

- Outlook has three distinct values: overcast, rainy, and sunny. So, C = 3.

- There are 4 overcast; thus, N-overcast = 4.

- Similarly, N-rainy = 5 and N-sunny = 5.

The intrinsic information of attribute Outlook is then computed as

```
S-outlook
= -(4/14×Log2(4/14)+5/14×Log2(5/14)+5/14×Log2(5/14))
= 1.58
```

The final gain ratio of attribute Outlook = $(0.94-0.69)/1.58 = 0.16$.

As information gain is a good measure for node split in our decision tree example, we won't compute gain ratio for each attribute. Instead, the attribute with the largest information gain is picked to split the tree node. In this case, Outlook is selected to split the tree node into three child nodes based on the three values: overcast, rainy, and sunny.

Since the overcast node has entropy of 0.0, it must be a leaf node. Another two nodes, rainy node and sunny node, can be further divided based on either Temperature, Windy, or Humidity.

Learn Decision Trees Through Excel

The preceding computation can be very tedious and error prone. Let's use Excel to ease the process. Open Chapter10-1a.xlsx; there is only one worksheet named level-1 (level-1 represents that this is the first level of tree nodes). Data in Chapter10-1a.xlsx is exactly the same as Table 10-1 and is shown in Figure 10-2.

⬒	A	B	C	D	E
1	**Temperature**	**Humidity**	**Windy**	**Outlook**	**Play**
2	hot	high	FALSE	overcast	yes
3	cool	normal	TRUE	overcast	yes
4	mild	high	TRUE	overcast	yes
5	hot	normal	FALSE	overcast	yes
6	mild	high	FALSE	rainy	yes
7	cool	normal	FALSE	rainy	yes
8	cool	normal	TRUE	rainy	no
9	mild	normal	FALSE	rainy	yes
10	mild	high	TRUE	rainy	no
11	hot	high	FALSE	sunny	no
12	hot	high	TRUE	sunny	no
13	mild	high	FALSE	sunny	no
14	cool	normal	FALSE	sunny	yes
15	mild	normal	TRUE	sunny	yes

level-1   ⊕

Figure 10-2. *Golf data in Excel*

Learn Decision Trees Through Excel

We need to set up our data in a proper table so that we can autofill formulas. We have seen such table organizations in previous chapters.

1. Enter "Sample-size" in cell A16 and 14 in cell B16. 14 is the data size.

2. Enter "p*log(p)" in cell E17. p*log(p) represents $P_K\log_2(P_k)$ in Equation (10-1).

3. Enter "entropy" in cell F17.

4. Merge cells B18 and B19 together; enter "Play" inside the merged cell.

5. Enter "yes" in cell C18 and "no" in cell C19.

By now, part of our worksheet looks like Figure 10-3 except that our worksheet has no numbers in cell D18 or D19. Don't worry, we are going to compute them soon.

◢	A	B	C	D	E	F
14	cool	normal	FALSE	sunny	yes	
15	mild	normal	TRUE	sunny	yes	
16	Sample-size	14				
17					p*log(p)	entropy
18		Play	yes	9		
19			no	5		
20						

Figure 10-3. *Golf dataset and table setup for Play*

Follow these instructions to complete the computation of the entropy for the target variable Play:

6. In cell D18, enter the formula `=COUNTIFS($E$2:$E$15,C18)`. This counts how many *yes* there are for Play.

7. Autofill from cell D18 to cell D19. D19 counts how many *no* for Play.

8. In cell E18, enter the formula `=D18/$B$16*LOG(D18/$B$16,2)`. This formula computes $P_{yes} \times Log2(P_{yes})$.

9. Autofill from E18 to E19. E19 computes $P_{no} \times Log2(P_{no})$.

10. Merge cells F18 and F19; enter the formula `=-SUM(E18:E19)` inside the merged cell. The result is the entropy for the target variable Play.

Our worksheet looks like Figure 10-4.

◢	A	B	C	D	E	F
16	Sample-size	14				
17					p*log(p)	entropy
18		Play	yes	9	-0.40977638	0.94028596
19			no	5	-0.53050958	
20						

Figure 10-4. *Entropy of Play is computed*

It's time to set up proper tables for the four attributes. The trick is once the first formula is defined correctly, how to allow a vertical autofill to work on different attributes that are arranged in separate columns (see Figure 10-2). This may sound challenging, but the function INDEX can help out.

The function INDEX requires an array (it is a table indeed) as the first input parameter. If the second parameter (row) is 0, the function returns a column specified by the column number in the array. In the dataset, Temperature, Humidity, Windy, and Outlook are columns 1, 2, 3, and 4. The expression INDEX($A$2:$E$15,0,1) fetches the Temperature column, while INDEX($A$2:$E$15,0,4) fetches the Outlook column.

Follow these instructions to set up the helping tables for the four attributes:

11. Enter 1, 1, 1, 2, 2, 3, 3, 4, 4, 4 into cells A22:A31, respectively.

12. Enter "yes", "no", "p*log(p)-yes", "p*log(p)-no", "weighted", "entropy", and "info gain" in cells D21:J21, respectively. Column "weighted" stores the weighted entropy for each attribute value.

13. Merge cells B22:B24; enter "Temperature" in the merged cell.

14. Enter "hot", "mild", and "cool" in cells C22, C23, and C24, respectively.

Part of our worksheet looks like Figure 10-5. Observe when aligning cells A22:A24 to C22:C24, "hot", "mild", and "cool" all correspond to 1 as they all are values of attribute Temperature which is the first column in table A1:E15.

	A	B	C	D	E	F	G	H	I	J
18		Play	yes	9	-0.40977638	0.94028596				
19			no	5	-0.53050958					
20										
21				yes	no	p*log(p)-yes	p*log(p)-r	weighted	entropy	info gain
22	1		hot							
23	1	Temperature	mild							
24	1		cool							
25	2									
26	2									
27	3									
28	3									
29	4									
30	4									
31	4									

Figure 10-5. *Table setup in process*

15. Merge cells B25 and B26; enter "Humidity" in the merged cell.

16. Enter "high" and "normal" in cells C25 and C26, respectively.

17. Merge cells B27 and B28; enter "Windy" in the merged cell.

18. Enter "TRUE" inside cell C27 and "FALSE" inside cell C28.

19. Merge cells B29:B31; enter "Outlook" in the merged cell.

20. Enter "overcast", "rainy", and "sunny" in cells C29, C30, and C31, respectively.

    Part of our worksheet is comparable with Figure 10-6.

	A	B	C	D	E	F	G	H	I	J
16	Sample-size	14								
17					p*log(p)	entropy				
18		Play	yes	9	-0.40977638	0.94028596				
19			no	5	-0.53050958					
20										
21				yes	no	p*log(p)-yes	p*log(p)-no	weighted entropy		info gain
22	1		hot							
23	1	Temperature	mild							
24	1		cool							
25	2	Humidity	high							
26	2		normal							
27	3	Windy	TRUE							
28	3		FALSE							
29	4		overcast							
30	4	Outlook	rainy							
31	4		sunny							

Figure 10-6. *Set up all the tables*

With the helping tables ready, follow these instructions to compute the entropy and information gains:

21. Enter the following formula in cell D22:

    =COUNTIFS(INDEX($A$2:$E$15,0,$A22), $C22,$E$2:$E$15,D$21)

    Cell A22 references 1; therefore, INDEX($A$2:$E$15,0,$A22) fetches the Temperature column. This formula counts how many data points there are whose Temperature is "hot" and Play is "yes" at the same time.

22. Autofill from cell D22 to E22, and then autofill together to cells D31:E31. This is shown in Figure 10-7.

◢	A	B	C	D	E	F
20						
21				yes	no	p*log(p)-yes
22	1		hot	2	2	
23	1	Temperature	mild	4	2	
24	1		cool	3	1	
25	2	Humidity	high	3	4	
26	2		normal	6	1	
27	3	Windy	TRUE	3	3	
28	3		FALSE	6	2	
29	4		overcast	4	0	
30	4	Outlook	rainy	3	2	
31	4		sunny	2	3	
32						

Figure 10-7. *Autofill across attributes*

23. Enter the following formula in cell F22:

    =IFERROR(D22/SUM($D22:$E22)*LOG(D22/SUM($D22:$E22), 2), 0)

    It is possible that SUM($D22:$E22) returns 0, and therefore there will be a divided-by-zero error from the expression D22/SUM($D22:$E22). In addition, the log function cannot take 0 as an input. Thus, the function IFERROR is used here to catch such errors. If an error occurs, 0 is returned.

    This formula computes $P_{yes} \times Log2(P_{yes})$ for the attribute value "hot".

24. Autofill from cell F22 to cell G22, then autofill together to cells F31:G31. Note cell G22 computes $P_{no} \times Log2(P_{no})$ for the attribute value "hot".

25. In cell H22, enter the formula =-SUM($D22:$E22)/$B$16*(F22+G22). The calculated value is the weighted sub-entropy for the value "hot" of attribute Temperature.

171

26. Autofill from cell H22 to cell H31.

Part of our worksheet looks like Figure 10-8.

	A	B	C	D	E	F	G	H	I	J
21				yes	no	p*log(p)-yes	p*log(p)-r	weighted	entropy	info gain
22	1		hot	2	2	-0.5	-0.5	0.285714		
23	1	Temperature	mild	4	2	-0.389975	-0.5283	0.393555		
24	1		cool	3	1	-0.311278	-0.5	0.231794		
25	2	Humidity	high	3	4	-0.523882	-0.4613	0.492614		
26	2		normal	6	1	-0.190622	-0.4011	0.295836		
27	3	Windy	TRUE	3	3	-0.5	-0.5	0.428571		
28	3		FALSE	6	2	-0.311278	-0.5	0.463587		
29	4		overcast	4	0	0	0	0		
30	4	Outlook	rainy	3	2	-0.442179	-0.5288	0.346768		
31	4		sunny	2	3	-0.528771	-0.4422	0.346768		

Figure 10-8. *Individual entropy values computed*

27. In cell I22, enter the formula =SUMIFS(H$22:H$31,$A$22:A$31,A22); autofill from cell I22 to I31. The formula in cell I22 calculates the entropy for the attribute Temperature.

28. In cell J22, enter the formula =F$18-I22 to obtain information gain for attribute Temperature. Autofill from cell J22 to J31. Compare our result with Figure 10-9.

	D	E	F	G	H	I	J
21	yes	no	p*log(p)-yes	p*log(p)-r	weighted	entropy	info gain
22	2	2	-0.5	-0.5	0.285714	0.911063	0.029223
23	4	2	-0.389975	-0.52832	0.393555	0.911063	0.029223
24	3	1	-0.31127812	-0.5	0.231794	0.911063	0.029223
25	3	4	-0.52388247	-0.46135	0.492614	0.78845	0.151836
26	6	1	-0.19062208	-0.40105	0.295836	0.78845	0.151836
27	3	3	-0.5	-0.5	0.428571	0.892159	0.048127
28	6	2	-0.31127812	-0.5	0.463587	0.892159	0.048127
29	4	0	0	0	0	0.693536	0.24675
30	3	2	-0.44217936	-0.52877	0.346768	0.693536	0.24675
31	2	3	-0.52877124	-0.44218	0.346768	0.693536	0.24675

Figure 10-9. *Entropy and information gain computed*

29. Merge cells I22:I24, I25:I26, I27:I28, I29:I31, J22:J24, J25:J26, J27:J28, and J29:J31, respectively. The computation is completed as shown in Figure 10-10. Since the attribute Outlook has the largest information gain, it is selected to split the level-1 tree node.

	C	D	E	F	G	H	I	J
18	yes		9	-0.40977638	0.94028596			
19	no		5	-0.53050958				
20								
21		yes	no	p*log(p)-yes	p*log(p)-n	weighted	entropy	info gain
22	hot	2	2	-0.5	-0.5	0.285714		
23	mild	4	2	-0.389975	-0.52832	0.393555	0.911063	0.029223
24	cool	3	1	-0.31127812	-0.5	0.231794		
25	high	3	4	-0.52388247	-0.46135	0.492614	0.78845	0.151836
26	normal	6	1	-0.19062208	-0.40105	0.295836		
27	TRUE	3	3	-0.5	-0.5	0.428571	0.892159	0.048127
28	FALSE	6	2	-0.31127812	-0.5	0.463587		
29	overcast	4	0	0	0	0		
30	rainy	3	2	-0.44217936	-0.52877	0.346768	0.693536	0.24675
31	sunny	2	3	-0.52877124	-0.44218	0.346768		

Figure 10-10. Level-1 split will be based on Outlook

30. We can draw a simple "tree chart" as shown in Figure 10-11. As the weighted entropy H-outlook-overcast is zero, the child node overcast(4,0) is a leaf node. The next task is to split the nodes rainy(3,2) and sunny(2,3).

	C	D	E	F	G	H	I	
25	high	3	4	-0.523882466	-0.46135	0.492614068	0.78845	0
26	normal	6	1	-0.190622075	-0.40105	0.295836389		
27	TRUE	3	3	-0.5	-0.5	0.428571429	0.892159	0
28	FALSE	6	2	-0.311278124	-0.5	0.4635875		
29	overcast	4	0	0	0	0		
30	rainy	3	2	-0.442179356	-0.52877	0.346768069	0.693536	(
31	sunny	2	3	-0.528771238	-0.44218	0.346768069		
32								
33								
34				Outlook				
35								
36		rainy(3,2)		overcast(4,0)		sunny(2,3)		
37								
38								

Figure 10-11. *A simple tree chart*

Follow these instructions to split the node rainy(3,2):

31. Make a copy of the worksheet level-1; rename the new worksheet level-2-rainy.

32. In the worksheet level-2-rainy, enter "rainy" in cell F1.

33. In cell B16, enter the formula =COUNTIFS($D$2:$D$15,$F$1). This will only count those cells whose Outlook value is "rainy".

Make sure that our worksheet looks exactly like Figure 10-12.

Continue as instructed in the following to complete the task:

◢	A	B	C	D	E	F
1	Temperature	Humidity	Windy	Outlook	Play	rainy
2	hot	high	FALSE	overcast	yes	
3	cool	normal	TRUE	overcast	yes	
4	mild	high	TRUE	overcast	yes	
5	hot	normal	FALSE	overcast	yes	
6	mild	high	FALSE	rainy	yes	
7	cool	normal	FALSE	rainy	yes	
8	cool	normal	TRUE	rainy	no	
9	mild	normal	FALSE	rainy	yes	
10	mild	high	TRUE	rainy	no	
11	hot	high	FALSE	sunny	no	
12	hot	high	TRUE	sunny	no	
13	mild	high	FALSE	sunny	no	
14	cool	normal	FALSE	sunny	yes	
15	mild	normal	TRUE	sunny	yes	
16	Sample-size	5				

Figure 10-12. *Working on splitting tree node rainy(3,2)*

34. Cell D18 has the formula =COUNTIFS($E$2:$E$15,C18). Insert
    ",$D$2:$D$15,$F$1" right after "C18" in the formula so that the
    formula becomes

    =COUNTIFS($E$2:$E$15,C18,$D$2:$D$15,$F$1)

    Again, this will only include those cells whose Outlook value is
    "rainy".

35. Autofill from cell D18 to cell D19.

36. Insert ",$D$2:$D$15,$F$1" into the formula of cell D22; make
    sure its formula becomes

    =COUNTIFS(INDEX($A$2:$E$15,0,$A22),$C22, $E$2:$E$15,D$21,$D$2:$D
    $15,$F$1)

37.   Autofill from D22 to E22, then autofill together to cells D31:E31.

Part of our worksheet should look like Figure 10-13.

	A	B	C	D	E	F	G	H	I	J
16	Sample-size		5							
17					p*log(p)	entropy				
18		Play	yes	3	-0.44217936	0.970950594				
19			no	2	-0.52877124					
20										
21		1		yes	no	p*log(p)-yes	p*log(p)-r	weighted	entropy	info gain
22		1		hot	0	0	0	0	0	
23		1	Temperature	mild	2	1	-0.389975	-0.52832	0.5509775	0.950978	0.019973
24		1		cool	1	1	-0.5	-0.5	0.4		
25		2		high	1	1	-0.5	-0.5	0.4	0.950978	0.019973
26		2	Humidity	normal	2	1	-0.389975	-0.52832	0.5509775		
27		3		TRUE	0	2	0	0	0	0	0.970951
28		3	Windy	FALSE	3	0	0	0	0		
29		4		overcast	0	0	0	0	0		
30		4	Outlook	rainy	3	2	-0.442179356	-0.52877	0.970950594	0.970951	0
31		4		sunny	0	0	0	0	0		

Figure 10-13. *All entropy and information gain values computed for rainy(3,2)*

38.   It seems that the node rainy(3,2) should be branched based on
Windy. As the two child nodes Windy-t(0,2) and Windy-f(3,0) both
have zero entropy, they both are leaf nodes. We can modify the
existing "chart" as Figure 10-14.

	B	C	D	E	F	G	H
33							
34					Outlook		
35							
36			rainy(3,2)		overcast(4,0)		sunny(2,3)
37							
38		Windy-t(0,2	Windy-f(3,0)				
39							

Figure 10-14. *Split node rainy(3,2) on Windy*

To split the node sunny(2,3) is fairly easy. Follow these
instructions:

39.   Make a copy of the worksheet level-2-rainy; rename the new
worksheet level-2-sunny.

40.   In worksheet level-2-sunny, change the text inside cell F1 to "sunny".

That's it; all the computations are automatically accomplished by Excel. The result looks like Figure 10-15.

◢	A	B	C	D	E	F	G	H	I	J
16	Sample-size	5								
17					p*log(p)	entropy				
18		Play	yes		2	-0.5287712	0.97095059			
19			no		3	-0.4421794				
20										
21	1			yes	no	p*log(p)-yes	p*log(p)-i	weighted	entropy	info gain
22	1		hot	0	2	0	0	0		
23	1	Temperature	mild	1	1	-0.5	-0.5	0.4	0.4	0.57095
24	1		cool	1	0	0	0	0		
25	2	Humidity	high	0	3	0	0	0	0	0.97095
26	2		normal	2	0	0	0	0		
27	3	Windy	TRUE	1	1	-0.5	-0.5	0.4	0.95098	0.01997
28	3		FALSE	1	2	-0.5283208	-0.39	0.5509775		
29	4		overcast	0	0	0	0	0		
30	4	Outlook	rainy	0	0	0	0	0	0.97095	0
31	4		sunny	2	3	-0.5287712	-0.4422	0.970950594		

Figure 10-15. *All entropy and information gain values computed for sunny(2,3)*

From the results shown in worksheet level-2-sunny, we notice that the tree node sunny(2,3) should be branched based on Humidity. As the two child nodes generated both have zero entropy, they both are leaf nodes. Thus, there is no need to split the tree anymore. The result is shown in Figure 10-16.

◢	B	C	D	E	F	G	H	I
32								
33								
34					Outlook			
35								
36			rainy(3,2)		overcast(4,0)		sunny(2,3)	
37								
38		Windy-t(0,2)	Windy-f(3,0)				Humidity-h(0,3)	Humidity-n(2,0)
39								
40								

◄   ►   ... | level-2-rainy | **level-2-sunny** | ⊕ | ⋮ ◄ |

Figure 10-16. *Decision tree construction completed*

The attribute Temperature has never been used to split the tree. This happens to be the case with this dataset, that is, the decision is not dependent on Temperature at all.

The complete result of the preceding process is available in the file Chapter10-1b.xlsx.

You may wonder if we can draw the tree chart by applying the same skills learned in the chapter about hierarchical clustering when we drew a dendrogram. Certainly we can. Again, it is quite some manual work and therefore I won't demonstrate it here but leave it up to you.

A Better Approach

Remember when we begin to split the node rainy(3,2) from step 31, we need to modify some formulas in the worksheet level-2-rainy? There is another approach that is very similar but does not require any formula modification. This approach is presented in the file Chapter10-2b.xlsx. This approach is more flexible and better, but the key formulas are a little bit more complicated. That is one reason why I didn't introduce it in the beginning. In addition, introducing this approach after we have had a clear mind about decision tree analysis in Excel can make us appreciate it more. Follow these instructions:

1. Open Chapter10-2a.xlsx. There is only one worksheet named "level-1" which is the same worksheet level-1 in the file Chapter10-1b.xlsx (the one we have worked on before). This worksheet looks like Figure 10-17.

▲	A	B	C	D	E	F	G	H	I	J
1	Temperature	Humidity	Windy	Outlook	Play					
2	hot	high	FALSE	overcast	yes					
3	cool	normal	TRUE	overcast	yes					
4	mild	high	TRUE	overcast	yes					
5	hot	normal	FALSE	overcast	yes					
6	mild	high	FALSE	rainy	yes					
7	cool	normal	FALSE	rainy	yes					
8	cool	normal	TRUE	rainy	no					
9	mild	normal	FALSE	rainy	yes					
10	mild	high	TRUE	rainy	no					
11	hot	high	FALSE	sunny	no					
12	hot	high	TRUE	sunny	no					
13	mild	high	FALSE	sunny	no					
14	cool	normal	FALSE	sunny	yes					
15	mild	normal	TRUE	sunny	yes					
16	Sample-size	14								
17					p*log(p)	entropy				
18		Play	yes	9	-0.40977638	0.940285959				
19			no	5	-0.53050958					
20										
21	1		yes	no	p*log(p)-yes	p*log(p)-r	weighted	entropy	info gain	
22	1	hot	2	2	-0.5	-0.5	0.285714286			

Figure 10-17. *A glance at worksheet level-1*

2. Enter the texts Temperature, Humidity, Windy, and Outlook in cells F1, G1, H1, and I1, respectively. Enter "<>" (without quotation marks) in cells F2:I2. "<>" means "not equal" in Excel. Refer to Figure 10-18.

▲	A	B	C	D	E	F	G	H	I
1	Temperature	Humidity	Windy	Outlook	Play	Temperature	Humidity	Windy	Outlook
2	hot	high	FALSE	overcast	yes	<>	<>	<>	<>
3	cool	normal	TRUE	overcast	yes				
4	mild	high	TRUE	overcast	yes				
5	hot	normal	FALSE	overcast	yes				
6	mild	high	FALSE	rainy	yes				
7	cool	normal	FALSE	rainy	yes				
8	cool	normal	TRUE	rainy	no				
9	mild	normal	FALSE	rainy	yes				
10	mild	high	TRUE	rainy	no				
11	hot	high	FALSE	sunny	no				
12	hot	high	TRUE	sunny	no				
13	mild	high	FALSE	sunny	no				
14	cool	normal	FALSE	sunny	yes				
15	mild	normal	TRUE	sunny	yes				

Figure 10-18. *Additional setup of the data table*

3. Enter the following formula in cell B16:

    ```
    =COUNTIFS(A2:A15,F2,B2:B15, G2,C2:C15,H2,D2:D15,I2)
    ```

    As cells F2:I2 all contain "<>", the preceding formula does not really set up any criteria for the function COUNTIFS. The value inside cell B16 is still 14.

4. In cell D18, revise the formula to

    ```
    =COUNTIFS($E$2:$E$15,C18,$A$2:$A$15,F$2, $B$2:$B$15,G$2,
    $C$2:$C$15,H$2,$D$2:$D$15,I$2)
    ```

    This formula takes the criteria in cells F2:I2 into consideration.

5. Autofill from cell D18 to cell D19.

6. In cell D22, revise the formula to

    ```
    =COUNTIFS(INDEX($A$2:$E$15,0,$A22),$C22, $E$2:$E$15,D$21,$A$2:$A
    $15,$F$2,$B$2:$B$15, $G$2,$C$2:$C$15,$H$2,$D$2:$D$15,$I$2)
    ```

    This formula takes the criteria in cells F2:I2 into consideration, too.

7. Autofill from cell D22 to cell E22, and then autofill together cells D22:E22 to cells D31:E31.

    By this time, everything in this worksheet should automatically work out. Part of our worksheet looks exactly like Figure 10-19, which is the same as we had before.

	A	B	C	D	E	F	G	H	I	J
16	Sample-size	14								
17						p*log(p)	entropy			
18		Play	yes	9	-0.40977638		0.940285959			
19			no	5	-0.53050958					
20										
21	1			yes	no	p*log(p)-yes	p*log(p)-r	weighted	entropy	info gain
22	1		hot	2	2	-0.5	-0.5	0.285714286		
23	1	Temperature	mild	4	2	-0.389975	-0.52832	0.393555357	0.911063	0.029223
24	1		cool	3	1	-0.311278124	-0.5	0.23179375		
25	2	Humidity	high	3	4	-0.523882466	-0.46135	0.492614068	0.78845	0.151836
26	2		normal	6	1	-0.190622075	-0.40105	0.295836389		
27	3	Windy	TRUE	3	3	-0.5	-0.5	0.428571429	0.892159	0.048127
28	3		FALSE	6	2	-0.311278124	-0.5	0.4635875		
29	4		overcast	4	0	0	0	0		
30	4	Outlook	rainy	3	2	-0.442179356	-0.52877	0.346768069	0.693536	0.24675
31	4		sunny	2	3	-0.528771238	-0.44218	0.346768069		

Figure 10-19. *The computation results of the better approach*

8. We are going to split the node rainy(3, 2) again. Make a copy of the worksheet level-1 and rename the new worksheet level-2-rainy as we did before. In cell I2 of worksheet level-2-rainy, enter "rainy" as shown in Figure 10-20. That's it! All computations in this worksheet are automatically completed by Excel.

	A	B	C	D	E	F	G	H	I
1	Temperature	Humidity	Windy	Outlook	Play	Temperature	Humidity	Windy	Outlook
2	hot	high	FALSE	overcast	yes	◇	◇	◇	rainy
3	cool	normal	TRUE	overcast	yes				
4	mild	high	TRUE	overcast	yes				
5	hot	normal	FALSE	overcast	yes				
6	mild	high	FALSE	rainy	yes				
7	cool	normal	FALSE	rainy	yes				
8	cool	normal	TRUE	rainy	no				
9	mild	normal	FALSE	rainy	yes				
10	mild	high	TRUE	rainy	no				
11	hot	high	FALSE	sunny	no				
12	hot	high	TRUE	sunny	no				
13	mild	high	FALSE	sunny	no				
14	cool	normal	FALSE	sunny	yes				
15	mild	normal	TRUE	sunny	yes				

Figure 10-20. *A glance at worksheet level-2-rainy*

9. To split on node sunny(2, 3), make a copy of worksheet level-2-rainy and rename the new worksheet level-2-sunny. In the worksheet level-2-sunny, change the text inside cell I2 to "sunny" as shown in Figure 10-21. Again, this is it. All computations are automatically completed by Excel for us.

	A	B	C	D	E	F	G	H	I
1	Temperature	Humidity	Windy	Outlook	Play	Temperature	Humidity	Windy	Outlook
2	hot	high	FALSE	overcast	yes	◇	◇	◇	sunny
3	cool	normal	TRUE	overcast	yes				
4	mild	high	TRUE	overcast	yes				
5	hot	normal	FALSE	overcast	yes				
6	mild	high	FALSE	rainy	yes				
7	cool	normal	FALSE	rainy	yes				
8	cool	normal	TRUE	rainy	no				
9	mild	normal	FALSE	rainy	yes				
10	mild	high	TRUE	rainy	no				
11	hot	high	FALSE	sunny	no				
12	hot	high	TRUE	sunny	no				
13	mild	high	FALSE	sunny	no				
14	cool	normal	FALSE	sunny	yes				
15	mild	normal	TRUE	sunny	yes				

Figure 10-21. *A glance at worksheet level-2-sunny*

Apply the Model

Decision trees are both classification and predictive models. We build the decision tree model to predict the class of future events. Open Chapter10-3a.xlsx; the data table looks like Figure 10-22.

◢	A	B	C	D	E	F	G
1	**Temperature**	**Humidity**	**Windy**	**Outlook**	**Play**	**Probability**	
2	mild	high	TRUE	overcast			
3	mild	normal	TRUE	sunny			
4	cool	high	TRUE	rainy			
5	cool	high	TRUE	rainy			
6	hot	high	FALSE	sunny			
7	hot	normal	TRUE	overcast			
8	mild	high	TRUE	sunny			
9	cool	high	TRUE	rainy			
10	cool	normal	TRUE	rainy			
11	mild	high	TRUE	rainy			
12	cool	high	FALSE	sunny			
13	cool	normal	FALSE	sunny			
14	hot	high	FALSE	overcast			
15	mild	high	FALSE	overcast			
16							
17				Outlook			
18							
19		rainy(3,2)		overcast(4,0)		sunny(2,3)	
20							
21	Windy-t(0,2)	Windy-f(3,0)				Humidity-h(0,3)	Humidity-n(2,0)

Figure 10-22. *Prediction for future events based on the decision tree model*

The data in the table are randomly generated. Our job is to predict the decision for Play based on the decision tree model which is also shown in Figure 10-22. This decision tree is a rule-based model, which can be best programmed using the IF function.

1. Enter the following formula in cell E2 and autofill from cell E2 to cell E15:

```
=IF(D2="overcast","yes", IF(D2="rainy",IF(C2="TRUE","no","yes"),
IF(B2="high","no","yes")))
```

This formula implements our simple decision tree rules.

2. Enter the following formula in cell F2:

```
=IF(D2="overcast",4/4, IF(D2="rainy",IF(C2="TRUE",2/2,3/3),
IF(B2="high",3/3,2/2)))
```

183

This formula computes the probability for each Play class. The probability of each Play class is computed baséd on the numbers inside each leaf node. For example, if the Outlook is overcast, the leaf node has 4 yes and 0 no; thus, the probability is 4/4 = 1. Given that the Outlook is sunny and Humidity is high, since the corresponding leaf node has 0 yes and 3 no, the probability for "no" is 3/3 = 1. Our data size is very small, which simplifies each leaf node to have data of only one class. This is the reason why all probability values are 1 as shown in Figure 10-23.

Note, we are using actual numbers instead of cell references in the preceding formula. This is for better demonstration.

	A	B	C	D	E	F
1	**Temperature**	**Humidity**	**Windy**	**Outlook**	**Play**	**Probability**
2	mild	high	TRUE	overcast	yes	1
3	mild	normal	TRUE	sunny	yes	1
4	cool	high	TRUE	rainy	no	1
5	cool	high	TRUE	rainy	no	1
6	hot	high	FALSE	sunny	no	1
7	hot	normal	TRUE	overcast	yes	1
8	mild	high	TRUE	sunny	no	1
9	cool	high	TRUE	rainy	no	1
10	cool	normal	TRUE	rainy	no	1
11	mild	high	TRUE	rainy	no	1
12	cool	high	FALSE	sunny	no	1
13	cool	normal	FALSE	sunny	yes	1
14	hot	high	FALSE	overcast	yes	1
15	mild	high	FALSE	overcast	yes	1

Figure 10-23. *Decision tree probability computed*

Computing the probability only based on the numbers inside each leaf node is debatable, in my opinion. However, this is not a topic of discussion in this book. The complete prediction results can be found in the file Chapter10-3b.xlsx which contains an additional worksheet that randomizes the input data for the demonstration purpose.

This wraps up another chapter. A critical part of decision tree method is to visualize the model. Certainly, Excel is not suitable to draw customized decision tree graphs. Nevertheless, this chapter demonstrates that Excel is capable of conducting decision tree analysis.

Reinforcement Exercises

Though the dataset used previously is good for demonstration, it is quite small and not impractical for real applications. For this chapter's reinforcement exercise, we are going to make use of the Car Evaluation dataset again (Bohanec, Marko. 1997. Car Evaluation. UCI Machine Learning Repository. https://doi.org/10.24432/C5JP48). Please open the file Chapter10-HW.xlsx to practice Decision Trees data mining method. You can reference to Chapter10-HW-withAnswers.xlsx whenever you need help.

Review Points

1. Entropy, information gain, and gain ratio

2. Set up helping tables

3. Functions IF, COUNTIFS, SUMIFS, and INDEX

4. Functions LOG and IFERROR

5. The operator <>

6. Copy a worksheet to continue a tree node split

EDA, Data Cleaning, and Feature Selection

Exploratory data analysis (EDA) and data cleaning are usually the first two topics in a traditional data mining course. So, it is legitimate to ask why I place them in Chapter 11.

When I started to self-learn data mining, the first book I read did place these two topics in the first three chapters. At that point of time, I had no idea how data mining methods work and was eager to learn the magic. I was not happy that after the first three chapters, I still had no idea what typical data mining methods were, let alone how they worked. It is after I had finished the whole book, I began to appreciate the necessity of EDA and data cleaning. Later, through my own practices and research in machine learning, I realized the importance of feature selection. Now you have accumulated enough knowledge about some data mining methods, it is time to develop some understanding about EDA, data cleaning, and feature selection, which are the three tasks we should perform before we construct our data mining models.

Please download the sample Excel files from `https://github.com/hhohho/Learn-Data-Mining-through-Excel-2` for this chapter's exercises.

Learn Exploratory Data Analysis Through Excel

A typical full data mining project starts with business understanding and data understanding because we should have a clear mind of what problems we try to solve and what data we need. The next phase is data preparation during which we collect and prepare the data. It is in the data preparation phase that we conduct exploratory data analysis and data cleaning tasks.

The keyword "exploratory" indicates that EDA refers to the initial investigations so that we understand the following regarding the data:

- Number of records

- Number of attributes

- Data type of each attribute

- Sum, mean, median, and variances of each numerical attribute

- Value range of each numerical attribute

- Unique values of categorical target and attributes

- Class distribution of the target

- If there are any missing data

- Data anomalies, etc.

Excel is fairly suitable for EDA because the data in a worksheet is directly visible and Excel has many functions and built-in features that allow easy completion of summary statistics operations and anomaly detection. Please open Chapter11-1a.xlsx which shows a dataset contributed by Chiharu Sano to UCI Machine Learning Repository. You shall notice that in the worksheet named "crx", there are 15 attributes and one target variable as shown in Figure 11-1.

	A	B	C	D	E	F	G	H	I	J	K	L	M	N	O	P
1	A1	A2	A3	A4	A5	A6	A7	A8	A9	A10	A11	A12	A13	A14	A15	Target
2	b	30.83	0	u	g	w	v	1.25	t	t	1	f	g	202	0	+
3	a	58.67	4.46	u	g	q	h	3.04	t	t	6	f	g	43	560	+
4	a	24.5	0.5	u	g	q	h	1.5	t	f	0	f	g	280	824	+
5	b	27.83	1.54	u	g	w	v	3.75	t	t	5	t	g	100	3	+
6	b	20.17	5.625	u	g	w	v	1.71	t	f	0	f	s	120	0	+
7	b	32.08	4	u	g	m	v	2.5	t	f	0	t	g	360	0	+
8	b	33.17	1.04	u	g	r	h	6.5	t	f	0	t	g	164	31285	+
9	a	22.92	11.585	u	g	cc	v	0.04	t	f	0	f	g	80	1349	+
10	b	54.42	0.5	y	p	k	h	3.96	t	f	0	f	g	180	314	+
11	b	42.5	4.915	y	p	w	v	3.165	t	f	0	t	g	52	1442	+
12	b	22.08	0.83	u	g	c	h	2.165	f	f	0	t	g	128	0	+
13	b	29.92	1.835	u	g	c	h	4.335	t	f	0	f	g	260	200	+
14	a	38.25	6	u	g	k	v	1	t	f	0	t	g	0	0	+
15	b	48.08	6.04	u	g	k	v	0.04	f	f	0	f	g	0	2690	+
16	a	45.83	10.5	u	g	q	v	5	t	t	7	t	g	0	0	+

Figure 11-1. *Excel provides a direct visual view of the dataset*

Let's complete the previously listed exploratory actions one by one in this worksheet.

1. Number of records:

    Since the dataset is in Excel, you can easily scroll down to the bottom to find out that there are 691-1 = 690 records. In case the dataset size is too large, you can type the formula =COUNTA(P:P) - 1 in cell Q2 to obtain the number of records. The COUNTA function counts the number of cells that are not empty in a range. Since there might be empty cells inside attributes, counting the target values is a reliable way to confirm the number of records.

2. Number of attributes:

    As Excel allows us to view the data directly, we can easily tell the number of attributes of a dataset in an Excel worksheet. Should you want to make use of a formula, please type =COUNTA(A1:P1) in cell Q1. Thus, there are 15 attributes and one target (label).

3. Data type of each attribute:

    The Excel worksheet immediately tells us through our eyes that attributes A1, A4–A7, A9, A10, A12, and A13 and the target are categorical text data, while A2, A3, A8, A11, A14, and A15 are numerical data.

4. Sum, mean, median, and variances of each numerical attribute:

    Certainly we can apply the functions SUM, AVERAGE, and MEDIAN to columns A2, A3, A8, A11, A14, and A15 to obtain their sum, mean, and median. I don't need to demonstrate how to achieve them here.

    For the variances of these numerical attributes, we need to apply the function VAR.P, which calculates the population variance. The population variance is calculated as $\dfrac{\sum_{i=1}^{n}(x_i - \bar{x})^2}{n}$. Since there are only 690 records, we can scroll down to the bottom of the worksheet to enter the formula =IFERROR(VAR.P(A2:A691), "") in cell A692 and then autofill to cell O692. The IFERROR function is used to take care of the nonnumerical attributes. Note, if there

are a large number of records, we can certainly insert a row right
above the attribute names to calculate and store the variances.
At this point of time, part of our worksheet should look like
Figure 11-2. Note, the word "Variance" was added inside cell Q692
to mark the row 692.

	A	B	C	D	E	F	G	H	I	J	K	L	M	N	O	P	Q
1	A1	A2	A3	A4	A5	A6	A7	A8	A9	A10	A11	A12	A13	A14	A15	Target	16
2	b	30.83	0	u	g	w	v	1.25	t	t		1 f	g	202	0	+	692
3	a	58.67	4.46	u	g	q	h	3.04	t	t		6 f	g	43	560	+	
4	a	24.5	0.5	u	g	q	h	1.5	t	f		0 f	g	280	824	+	
5	b	27.83	1.54	u	g	w	v	3.75	t	t		5 t	g	100	3	+	
6	b	20.17	5.625	u	g	w	v	1.71	t	f		0 f	s	120	0	+	
7	b	32.08	4	u	g	m	v	2.5	t	f		0 t	g	360	0	+	
8	b	33.17	1.04	u	g	r	h	6.5	t	f		0 t	g	164	31285	+	
9	a	22.92	11.585	u	g	cc	v	0.04	t	f		0 f	g	80	1349	+	
10	b	54.42	0.5	y	p	k	h	3.96	t	f		0 f	g	180	314	+	
689	a	25.25	13.5	y	p	ff	ff	2	f	t		1 t	g	200	1	-	
690	b	17.92	0.205	u	g	aa	v	0.04	f	f		0 f	g	280	750	-	
691	b	35	3.375	u	g	c	h	8.29	f	f		0 t	g	0	0	-	
692		142.7796	24.74619					11.18292			23.61391			30164.17	27105828		Variance

Figure 11-2. *The variances of the numerical attributes are stored in row 692*

The variances are not computed on the same scale by default
and therefore you will notice that attribute A15 has the largest
variance. Sometimes the variances are computed after all the
attributes are normalized into the same range. I will address the
normalization issue in detail when I explain the feature selection.
Here, let's normalize all the numerical attributes into the range
of 0–100.

To compute the variances on the same scale of 0–100, let's make
a copy of the current worksheet and rename the new worksheet
crx2. In the cell A2 of worksheet crx2, enter the following formula:

```
=IFERROR((crx!A2-MIN(crx!A$2:A$691))/(MAX(crx!A$2:A$691)-
MIN(crx!A$2:A$691))*100, crx!A2)
```

This formula processes the corresponding data item inside
worksheet crx before placing it inside cell A2. The IFERROR
function guarantees if the column A data is not numerical, an
error will occur and therefore the data at crx!A2 will be placed
inside cell A2 without normalization. If the column A data is
numerical, it then normalizes the data value inside crx!A2 into the
range of 0–100 inclusively.

Autofill from A2 to O2 and then autofill together to A691:O691, we shall obtain the variances of the normalized data which illustrate that the largest variance is in attribute A2. Part of our crx2 worksheet should look like Figure 11-3.

	A	B	C	D	E	F	G	H	I	J	K	L	M	N	O	P	Q
1	A1	A2	A3	A4	A5	A6	A7	A8	A9	A10	A11	A12	A13	A14	A15	Target	16
2	b	25.68421	0	u	g	w	v	4.385965	t	t	1.492537	f	g	10.1	0	+	690
3	a	67.54887	15.92857	u	g	q	h	10.66667	t	t	8.955224	f	g	2.15	0.56	+	
4	a	16.16541	1.785714	u	g	q	h	5.263158	t	f	0	f	g	14	0.824	+	
5	b	21.17293	5.5	u	g	w	v	13.15789	t	t	7.462687	t	g	5	0.003	+	
6	b	9.654135	20.08929	u	g	w	v	6	t	f	0	f	s	6	0	+	
7	b	27.56391	14.28571	u	g	m	v	8.77193	t	f	0	t	g	18	0	+	
8	b	29.20301	3.714286	u	g	r	h	22.80702	t	f	0	f	g	8.2	31.285	+	
9	a	13.78947	41.375	u	g	cc	v	0.140351	t	f	0	f	g	4	1.349	+	
10	b	61.15789	1.785714	y	p	k	h	13.89474	t	f	0	f	g	9	0.314	+	
689	a	17.29323	48.21429	y	p	ff	ff	7.017544	f	t	1.492537	t	g	10	0.001	-	
690	b	6.270677	0.732143	u	g	aa	v	0.140351	f	f	0	f	g	14	0.75	-	
691	b	31.95489	12.05357	u	g	c	h	29.08772	f	f	0	t	g	0	0	-	
692		322.8664	315.6402					137.6783			52.60395			75.41043	27.10583		Variance

Figure 11-3. *The variances after numerical attributes are normalized into the range 0–100*

5. Value range of each numerical attribute:

   The value range usually indicates the largest and smallest values of a numerical attribute. We can apply the Excel function MAX and MIN to obtain the largest and smallest values of each numerical attribute. The value range can help identify outliers and data anomalies.

6. Unique values of categorical target and attributes:

   To obtain the unique values of categorical attributes, we can make use of the Excel built-in feature "Remove Duplicates". Before we remove duplicates for each categorical attribute, we'd better make a copy of the worksheet crx and rename the new worksheet crx3.

   Inside the crx3 worksheet, please click the column letter G ➤ click the Data tab ➤ select Remove Duplicates ➤ on the upcoming menu, select the option "Continue with the current selection" and then click the button "Remove Duplicates...". This is shown in Figure 11-4.

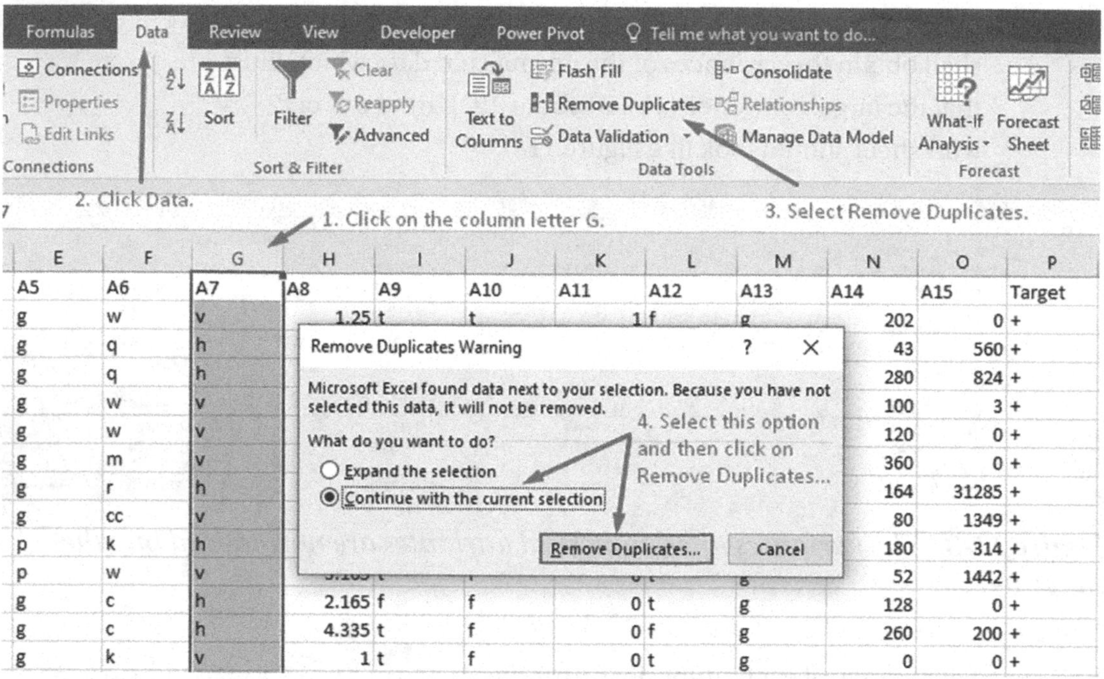

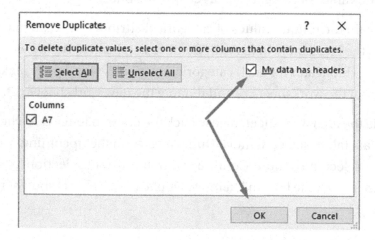

Figure 11-4. *Select Remove Duplicates*

Another small menu shows up. Follow the instruction in Figure 11-5 to click the OK button. Finally, a last small menu appears which tells us that there are "680 duplicates found and removed; 11 unique values remain."

Figure 11-5. *Make sure that "My data has headers" is checked*

These actions removed the duplicates for attribute A7. We need to repeat these actions on all other categorical attributes and the target.

Interestingly, you will realize that some attributes have the value "?" inside. Well, the question mark symbol "?" is commonly used to represent a missing value. So, those "?" in the crx3 worksheet indicates that there are missing values in the corresponding attributes. We will talk about missing values soon.

7. Class distribution of the target:

   Once we have the unique values of the target available in the worksheet crx3, we can easily acquire the class distribution of the target. Inside the crx3 worksheet, enter the formula =COUNTIFS(crx!P2:P691, P2) inside R2 and then autofill to R3. We can immediately tell that there are 307 and 383 data points for the two classes "+" and "-", respectively.

   A well-balanced dataset should have balanced classes, that is, all the classes should have about the same number of data points. In this specific case, the two classes weigh about 44.5% and 54.5%, which could be considered well balanced.

   What if a dataset has imbalanced classes? If something like this happens, you should try your best to increase the data points for the underrepresented class or classes. If the data size is large enough, you can also decrease the data points of the overrepresented class or classes. In the worst case when you cannot change the imbalance at all, you then should make sure that your machine learning model is sensitive enough on the underrepresented class or classes.

8. If there are any missing data:

   Missing values can exist in different forms, such as empty space or the question mark "?". The preceding action to remove duplicates helps us identify unusual categorical values which are missing values. For empty spaces, we can make use of the function COUNTBLANK to count them.

Go back to the worksheet crx, enter =COUNTBLANK(A2:A691) inside A693, and then autofill to P693. Fortunately, we do not have any empty spaces in our dataset.

To count how many missing values marked by "?" in each attribute, enter =COUNTIFS(A2:A691, "~?") inside A694 and then autofill to P694. Be aware that the question mark "?" is a special character in Excel. Therefore, to insert a literal "?" in this formula, we must prefix it with "~".

The complete result is in Chapter11-1b.xlsx. Observe that words "Variance", "Blank Cells", and "Missing Data" are inside cells Q692, Q693, and Q694. These words are for explanation.

9.   Data anomalies:

Data anomaly is different from outlier. Outliers are valid but their values are outside of the normal range. It is a difficult call if we should keep or drop the outlier data records.

Data anomaly references unexpected data or data inconsistence, either an undesired data type or a data value out of range. Generally speaking, when we collect data, we should be aware of the desired data type and data range for each attribute; otherwise, we won't have a standard to distinguish unexpected data values. For numerical data, we can use the IF function to identify any numerical value outside the desired range. For undesired data types, there are other functions to take care of them.

To practice data anomaly detection, open Chapter11-2a.xlsx in which I specifically created some data anomalies. Let's assume that we know the desired data types and ranges for each attribute. Specifically, the values must be >= 0 for numerical attributes.

Again, we will make a copy of the existing worksheet crx and rename the new worksheet crx2.

Inside worksheet crx2, enter =IF(ISNONTEXT(crx!A2), "N", "") in A2. Since we know that the data type must be alphabet for attribute A1, the function ISNONTEXT is used to determine if the corresponding value inside the worksheet crx is non-text. If it is, "N" is displayed; otherwise, nothing will be displayed in cell A2. Autofill from A2 to A691 by double-clicking on the lower right corner of the cell A2.

For attributes A2 and A3, we know that they must be numerical values and they cannot be less than 0. Thus, enter =IF(ISTEXT(crx!B2), "T", IF(crx!B2<0, "F", "")) in B2 and autofill to C2, then autofill together to B691:C691. We will immediately notice "F" in C6 and "T" in C8 because this formula returns "T" if crx!B2 contains text; else returns "F" if crx!B2 <0; else displays nothing in B2.

Enter =IF(ISNONTEXT(crx!D2), "N", "") in D2 and autofill to D691; we will also find "N" displayed in D15.

I hope you have realized the advantage of making copies of the original worksheet. Certainly, when there is "F", "T", or "N" showing up in cells C6, C8, and D15, respectively, inside the worksheet crx2, we immediately become knowledgeable that there are data anomalies at the exact same locations (C6, C8, and D15) inside the original worksheet crx.

At this point, part of our worksheet crx2 should look like Figure 11-6.

	A	B	C	D	E	F	G	H	I
1	A1	A2	A3	A4	A5	A6	A7	A8	A9
2					g	w	v	1.25	t
3					g	q	h	3.04	t
4					g	q	h	1.5	t
5					g	w	v	3.75	t
6			F		g	w	v	1.71	t
7					g	m	v	2.5	t
8			T		g	r	h	6.5	t
9					g	cc	v	0.04	t
10					p	k	h	3.96	t
11					p	w	v	3.165	t
12					g	c	h	2.165	f
13					g	c	h	4.335	t
14					g	k	v	1	t
15				N	g	k	v	0.04	f

Figure 11-6. *Cells C6, C8, and D15 display unexpected data values*

There are more unexpected data inside the worksheet. I would like to leave for you to finish the rest attributes. Note, you can certainly make use of different Excel functions to achieve the same purpose.

The completed result is available in Chapter11-2b.xlsx. Because the preceding anomaly detection process is pretty tedious, there is another file Chapter11-2c.xlsx in which only one formula is needed for all attributes. Please take a look.

Learn Data Cleaning Through Excel

Data cleaning usually applies on data records instead of data attributes, that is, on rows instead of columns. When there are missing data or unexpected data in a record, the safest approach is to delete this record since keeping it can introduce errors in our data mining result. The famous slogan "garbage in garbage out" says that if the data is not of high quality, the data mining result for sure won't be of high quality no matter whatever we do. Nevertheless, there are other data wrangling methods that a data engineer can adapt, though I usually do not recommend if your data size is large enough without those records.

A common technique is to replace a missing numerical data value with a constant (such as 0), or mean, median, or mode of the attribute. In certain cases, for example, if the unexpected value for a numerical data type is "high", we may consider replacing this unexpected value with a number larger than the mean or median.

For categorical attributes, a missing or null value can be replaced with a constant such as "missing".

When an attribute has too many low-quality data values, we should consider dropping this attribute as it is problematic by nature and becomes the data tumor. There is no one-for-all cutoff that determines when we should remove a column, it is upon experience and test and trial.

A more advanced approach is to develop a machine learning model to predict the missing or null values. For example, assume we have A1, A2, ..., A10 attributes and the label T, and there are some missing values inside A1. We can then develop a machine learning model that treats A1 as the target, but all other attributes include the label T as the predictors. The model is trained with the complete records and then used to predict for the missing A1 values.

Open Chapter11-3a.xlsx. You may realize it is the same as Chapter-2b.xlsx. The worksheet crx2 has already marked where the missing data or unexpected values are located. A quick looking through should let us be aware that there are not many data records to delete, compared to the data size. Thus, we should proceed to delete those records marked in worksheet crx2. We could start our operations from the worksheet crx2, or we can restart from the original worksheet crx.

Let's restart from the original worksheet crx by making a copy of it and rename the new worksheet crx3. Inside the worksheet crx3, please follow the succeeding instructions to complete our data cleaning effort. Note, for data cleaning, we must consider both missing data and data anomalies.

1. In cell A2, enter the following formula:

    ```
    =IF(COUNT(crx!A2:A691)/COUNTA(crx!A2:A691)>0.8, IF(ISTEXT(crx!A2),
    ROW(A2), IF(crx!A2<0, ROW(A2), "")), IF(OR(ISNONTEXT(crx!A2),
    crx!A2="?"), ROW(A2), ""))
    ```

    In this formula, `COUNT(crx!A2:A691)/COUNTA(crx!A2:A691)>0.8`
    is used to determine if an attribute is text or numerical data.
    Based on the data type, this attribute is examined for missing or
    unexpected data accordingly. Be advised that the function ISTEXT
    can detect the question mark "?".

Autofill from A2 to O691.

At this point, part of our worksheet should look like Figure 11-7.

	A	B	C	D	E	F	G	H	I	J	K	L	M	N	O	P
1	A1	A2	A3	A4	A5	A6	A7	A8	A9	A10	A11	A12	A13	A14	A15	Target
2																+
3																+
4																+
5												5		5		+
6			6													+
7															7	+
8			8												8	+
9																+
10																+
11																+
12																+
13																+
14																+
15				15												+

Figure 11-7. *Row numbers of the to-be-deleted records are marked*

2. Type "toKeep" inside cell Q1, enter =IF(SUM(A2:O2)>0, 1, 0) in Q2, and autofill to Q691. This formula determines if there is a row number inside A2:O2, then this row needs to be deleted by marking it as 1. This will help us later to keep those wanted records.

3. Again, let's make a copy of the worksheet crx so that we do not directly delete records inside the original worksheet. This is very important in data cleaning. Rename the new worksheet crx4.

4. Copy the cells Q1:Q691 in worksheet crx3 and paste them by values into the same Q1:Q691 region in worksheet crx4. Note, the paste action must be by values only. Part of our worksheet crx4 should look like Figure 11-8.

	A	B	C	D	E	F	G	H	I	J	K	L	M	N	O	P	Q
1	A1	A2	A3	A4	A5	A6	A7	A8	A9	A10	A11	A12	A13	A14	A15	Target	toKeep
2	b	30.83	0	u	g	w	v	1.25	t	t	1	f	g	202	0	+	0
3	a	58.67	4.46	u	g	q	h	3.04	t	t	6	f	g	43	560	+	0
4	a	24.5	0.5	u	g	q	h	1.5	t	f	0	f	g	280	824	+	0
5	b	27.83	1.54	u	g	w	v	3.75	t	t	5	-9	g	-3	3	+	1
6	b	20.17	-6.5	u	g	w	v	1.71	t	f	0	f	s	120	0	+	1
7	b	32.08	4	u	g	m	v	2.5	t	f	0	t	g	360	x	+	1
8	b	33.17	?	u	g	r	h	6.5	t	f	0	t	g	164	-1	+	1
9	a	22.92	11.585	u	g	cc	v	0.04	t	f	0	f	g	80	1349	+	0
10	b	54.42	0.5	y	p	k	h	3.96	t	f	0	f	g	180	314	+	0
11	b	42.5	4.915	y	p	w	v	3.165	t	f	0	t	g	52	1442	+	0
12	b	22.08	0.83	u	g	c	h	2.165	f	f	0	t	g	128	0	+	0
13	b	29.92	1.835	u	g	c	h	4.335	t	f	0	f	g	260	200	+	0
14	a	38.25	6	u	g	k	v	1	t	f	0	t	g	0	0	+	0
15	b	48.08	6.04	34.6	g	k	v	0.04	f	f	0	f	g	0	2690	+	1

Figure 11-8. *The worksheet crx4*

5. Let's create a blank new worksheet and name it "cleaned-crx". We are going to store our cleaned data inside this worksheet later.

6. Go back to the worksheet crx4. Select all the columns A–Q ➤ click the Data tab ➤ click Filter ➤ click the arrow inside the cell Q1 ➤ select Number Filters ➤ select the option Equals.... This is illustrated by Figure 11-9.

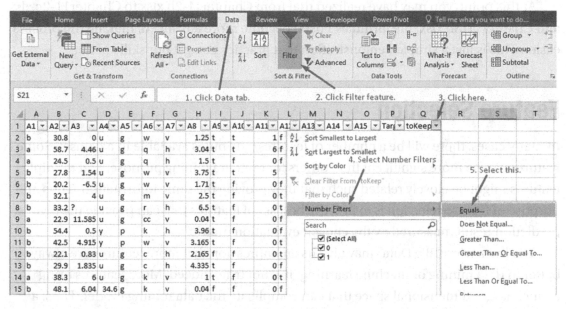

Figure 11-9. *Prepare to filter records based on the column "toKeep"*

7. A Custom AutoFilter window shows up. Follow the instructions in Figure 11-10 to proceed.

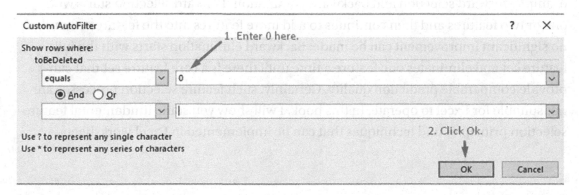

Figure 11-10. *Set up Custom AutoFilter*

8. Now in the worksheet crx4, only those to-be-kept records show
    up. Select all the data and copy/paste them into the cell A1 of the
    worksheet "cleaned-crx". That's it! The worksheet cleaned-crx
    contains the cleaned data.

The completed work is contained in the worksheet Chapter11-3b.xlsx. Note, you can
delete column Q inside the worksheet cleaned-crx.

At this point, you may have realized that from Chapter11-2b.xlsx to Chapter11-2c.xlsx
and Chapter11-3b.xlsx, the formulas I used become more advanced, making the process
easier. This is just a simple demonstration that practice can get our skills better.

Feature Selection

In many cases, there will be a number of features/attributes available in a dataset. Some
features can be more significant than others for our data mining purpose, and multiple
features might be closely related such that they provide the same information. Thus, the
purpose of feature selection is to find the best set of features that can provide the highest
prediction result measured by the chosen evaluation metric.

The buzz word "Big Data" may give us an impression that more features are always
better in data mining or machine learning. It is not true, indeed. More features bring
a much larger dimensional space that can complicate the data mining model. Thus, a
critical skill in data mining is to discover the most determinant features or attributes.
For example, what are the most important several features that determine the customer
attitudes toward a credit card offer?

In machine learning, there are quite some techniques for feature selections, for
example, forward selection and backward elimination. Forward selection starts with
one or two features and then continues to add more features into the feature set until
no significant improvement can be made. Backward elimination starts with the full
feature set and eliminates one feature a time until there is a core feature set that can
provide comparable prediction quality. Certainly, such feature selection processes are
not suitable for Excel to operate. In this book, I will show you some fundamental feature
selection principles and techniques that can be implemented in Excel worksheets.

There are three general principles regarding feature selections:

1. Features on the same scale work better.

2. Features with larger variances contribute more.

3. Related features tell the same data story.

As you can see, the first principle is to select features on the same scale. But what if none of the numerical features are on the same scale? This is something I promised to discuss in detail.

Normalization and Standardization

To make all numerical attributes on the same scale, there are two commonly used methods: normalization and standardization.

Standardization is the same as computing the Z-score of each value, that is, each value of an attribute is transformed to its corresponding Z-score by Equation (11-1):

$$z = \frac{x - \mu}{\sigma} \tag{11-1}$$

where x is a single value of an attribute, μ is the mean of the attribute, and σ is the standard deviation of this attribute.

Normalization, the traditional normalization method, is to scale up or down all the numerical values into the same range, commonly 0–1 or 0–100.

Both methods can be used to transform numerical attributes into the same scale. However, they both have their own advantages and disadvantages. Moreover, I personally believe that another normalization method, normalization by median, is an approach more appropriate than both the traditional normalization method and standardization method when we need to preserve the variances. Normalization by median is commonly used in bioinformatics study of gene expression. To compare these three methods, please open the Excel file Chapter11-scale-a.xlsx. Immediately, Figure 11-11 will be reflected in your eyes.

◢	A	B	C	D	E	F	G	H	I	J	K	L	M
1			Original			Normalization			Standardization			Normaization by Median	
2		A1	A2	A3	A1	A2	A3	A1	A2	A3	A1	A2	A3
3		1	100	5.01									
4		2	200	5.02									
5		3	300	5.03									
6		4	400	5.04									
7		5	500	5.05									
8		6	600	5.06									
9		7	700	5.07									
10		8	800	5.08									
11		9	900	5.09									
12		10	1000	5.1									
13	Variance												

Figure 11-11. *Three faked attributes A1, A2, and A3*

So far, we have had some experiences with the traditional normalization, which allows us able to tell that attribute A2 can be easily scaled down to the same as A1. What about the attribute A3? Well, let's normalize all the three attributes into the range 0–100 in the section of Normalization (columns E–G) first. In cell E3, enter the following formula and autofill from E3 to G3 then autofill together to E12:G12:

`=(B3-MIN(B$3:B$12))/(MAX(B$3:B$12)-MIN(B$3:B$12))*100`

Everything looks good. Now, let's apply Equation (11-1) to the three attributes by entering the following formula in H3 and autofill to J3 then autofill together to H12:J12:

`=(B3-AVERAGE(B$3:B$12))/STDEV.P(B$3:B$12)`

Here, we treat the ten records as the whole population and therefore apply the function STDEV.P on attribute A1. Observe that the values in H3:J12 have both negative and positive values.

Notice that attributes A1, A2, and A3 are transformed into the same values by the two previous methods. Something like this is unlikely to happen in reality. I designed them this way for demonstration purpose.

A good transformation should preserve the variances of the data. So, enter the formula =VAR.P(B3:B12) in B13 and then autofill to J13. At this point of time, our worksheet should look like Figure 11-12.

▲	A	B	C	D	E	F	G	H	I	J	K	L	M
1		Original			Normalization			Standardization			Normaization by Median		
2		A1	A2	A3	A1	A2	A3	A1	A2	A3	A1	A2	A3
3		1	100	5.01	0	0	0	-1.5667	-1.5667	-1.5667			
4		2	200	5.02	11.11111	11.11111	11.11111	-1.21854	-1.21854	-1.21854			
5		3	300	5.03	22.22222	22.22222	22.22222	-0.87039	-0.87039	-0.87039			
6		4	400	5.04	33.33333	33.33333	33.33333	-0.52223	-0.52223	-0.52223			
7		5	500	5.05	44.44444	44.44444	44.44444	-0.17408	-0.17408	-0.17408			
8		6	600	5.06	55.55556	55.55556	55.55556	0.174078	0.174078	0.174078			
9		7	700	5.07	66.66667	66.66667	66.66667	0.522233	0.522233	0.522233			
10		8	800	5.08	77.77778	77.77778	77.77778	0.870388	0.870388	0.870388			
11		9	900	5.09	88.88889	88.88889	88.88889	1.218544	1.218544	1.218544			
12		10	1000	5.1	100	100	100	1.566699	1.566699	1.566699			
13	Variance	8.25	82500	0.000825	1018.519	1018.519	1018.519	1	1	1			

Figure 11-12. *Both normalization and standard fail to preserve certain attributes' variances*

Figure 11-12 shows that both normalization and standardization render the three attributes the same variance. Since A2 = 100 x A1, we may agree that A1 and A2 should have the same variance after either normalization or standardization. But the original A3 seems to have little variance which was normalized/standardized to be the same as that of A1 or A3.

Normalization by median is a method that can scale the attribute into the same range and preserve the variance. It scales up or down attributes based on the medians. This method starts with selecting a reference attribute, for example, A1. Then it goes ahead to assume that all other attributes should have the same median as that of the reference attribute. Finally, it scales up or down an attribute based on the ratio of the two medians between this attribute and the reference attribute.

So, enter the following formula in K3, autofill to M3, and then autofill together to K12:M12:

```
=B3/(MEDIAN(B$3:B$12)/MEDIAN($B$3:$B$12))
```

Also, autofill from J13 to M13. Our worksheet should look like Figure 11-13.

◢	A	B	C	D	E	F	G	H	I	J	K	L	M
1			Original			Normalization			Standardization			Normaization by Median	
2		A1	A2	A3	A1	A2	A3	A1	A2	A3	A1	A2	A3
3		1	100	5.01	0	0	0	-1.5667	-1.5667	-1.5667	1	1	5.451039
4		2	200	5.02	11.11111	11.11111	11.11111	-1.21854	-1.21854	-1.21854	2	2	5.461919
5		3	300	5.03	22.22222	22.22222	22.22222	-0.87039	-0.87039	-0.87039	3	3	5.472799
6		4	400	5.04	33.33333	33.33333	33.33333	-0.52223	-0.52223	-0.52223	4	4	5.48368
7		5	500	5.05	44.44444	44.44444	44.44444	-0.17408	-0.17408	-0.17408	5	5	5.49456
8		6	600	5.06	55.55556	55.55556	55.55556	0.174078	0.174078	0.174078	6	6	5.50544
9		7	700	5.07	66.66667	66.66667	66.66667	0.522233	0.522233	0.522233	7	7	5.51632
10		8	800	5.08	77.77778	77.77778	77.77778	0.870388	0.870388	0.870388	8	8	5.527201
11		9	900	5.09	88.88889	88.88889	88.88889	1.218544	1.218544	1.218544	9	9	5.538081
12		10	1000	5.1	100	100	100	1.566699	1.566699	1.566699	10	10	5.548961
13	Variance	8.25	82500	0.000825	1018.519	1018.519	1018.519	1	1	1	8.25	8.25	0.000977

Figure 11-13. *The normalization by median method preserves attribute variances*

Figure 11-13 illustrates that the variance of A3 is also preserved through normalization by median.

In data mining projects, whether or not and how to normalize or standardize is always a puzzle requiring sophisticated assessment. In many cases, we can only be sure after multiple rounds of experiments.

Correlation

Correlation measures the linear association strength between two numerical variables or attributes. The strength is expressed by the correlation coefficient which must be in the range of -1 to 1 inclusively. Given two variables X and Y, if they are positively correlated, then X and Y go in the same direction. For example, if X is the daily temperature while Y is the ice cream sale. The higher the X, the larger the Y; or the lower the X, the smaller the Y. If X and Y are negatively correlated, then they go in the opposite directions. For example, the car mileage and car weight. When the correlation coefficient is close to zero, there is no correlation between X and Y. The computation of correlation coefficient is very simple in Excel by the function CORREL. This is shown in Figure 11-14.

▲	A	B	C	D	E
1	1	11		20	10
2	2	12		19	9
3	3	13		18	8
4	4	14		17	7
5	5	15		16	6
6	6	16		15	5
7	7	17		14	4
8	8	18		13	3
9	9	19		12	2
10	10	20		11	1
11					
12			=CORREL(A1:B10, D1:E10)		

Figure 11-14. *Correlation computation in Excel*

Learn Feature Selection Through Excel

Open the file Chapter11-4a.xlsx; you will find out there are two worksheets: "cleaned-crx" and "encoded". Worksheet cleaned-crx is the same as its counterpart in Chapter11-3b.xlsx. However, we are going to start our feature selection from the worksheet "encoded". If you compare the two worksheets, you will realize that worksheet encoded has an additional attribute named "Customer" in column A. This column is added to serve as the ID for each record. In addition, all the categorical attributes have been transformed into numerical attributes (integers) inside the worksheet encoded. Here, I performed an "encoding" action to make the data ready for feature selection operations.

Some data mining methods can only work with numerical values; thus, it is very common to transform categorical values into numerical, a process called "encoding". There are two types of commonly used encoding schemes: one hot encoding and ordinal encoding. Which one to use is based on the type of the categorical attribute. Generally speaking, all categorical attributes can be divided into two categories: nominal or ordinal. Nominal categorical attribute has no specific order, for example, the colors of eyes (yellow, black, brown, blue). Ordinal categorical attribute has specific order, for example, the grades A, B, C, D, F.

One hot encoding usually applies to nominal categorical attributes where a new attribute of values 0 or 1 is created for each category. For example, assume we have an attribute named EyeColor which has three categories: yellow, black, and blue. One hot encoding will then replace this EyeColor attribute with three new attributes, possibly named "yellow", "black", and "blue", one for an eye color. The 1s values in the new attribute "yellow" denote the presence of yellow eye color, while 0s values denote the absence of the yellow eye color. The same scheme applies to the "black" and "blue" attributes.

Ordinal encoding is much simpler. Given the grades A, B, C, D, F, we can simply transform them into 4–0, or 0–4. Because the sample dataset we are using does not provide any meaningful information about the categorical attributes, I applied ordinal encoding to all the categorical attributes.

Follow these instructions to practice feature selections in Excel.

1. Make a copy of the worksheet encoded and rename the new worksheet "normalized".

2. In cell B2 of the worksheet normalized, enter the following formula:

    ```
    =IFERROR((encoded!B2-MIN(encoded!B$2:B$647))/
    (MAX(encoded!B$2:B$647)-MIN(encoded!B$2:B$647))*100, encoded!B2)
    ```

    This formula normalizes the value in B2 of the attribute A1 into the 0–100 range.

3. Autofill from B2 to P2 then autofill from B2:P2 to B647:P647. Part of our normalized worksheet should look like Figure 11-15.

	A	B	C	D	E	F	G	H	I	J	K	L	M	N	O	P	Q
1	Customer	A1	A2	A3	A4	A5	A6	A7	A8	A9	A10	A11	A12	A13	A14	A15	Target
2	C1	100	27.11111	0	50	0	92.30769	87.5	4.385965	100	100	1.492537	0	0	10.1	0	1
3	C2	0	71.30159	15.92857	50	0	76.92308	37.5	10.66667	100	100	8.955224	0	0	2.15	0.56	1
4	C3	0	17.06349	1.785714	50	0	76.92308	37.5	5.263158	100	0	0	0	0	14	0.824	1
5	C4	0	14.55556	41.375	50	0	15.38462	87.5	0.140351	100	0	0	0	0	4	1.349	1
6	C5	100	64.55556	1.785714	100	100	61.53846	37.5	13.89474	100	0	0	0	0	9	0.314	1
7	C6	100	45.63492	17.55357	100	100	92.30769	87.5	11.10526	100	0	0	100	0	2.6	1.442	1
8	C7	100	13.22222	2.964286	50	0	7.692308	37.5	7.596491	0	0	0	100	0	6.4	0	1
9	C8	100	25.66667	6.553571	50	0	7.692308	37.5	15.21053	100	0	0	0	0	13	0.2	1
10	C9	0	38.88889	21.42857	50	0	61.53846	87.5	3.508772	100	0	0	100	0	0	0	1
11	C10	0	50.92063	37.5	50	0	76.92308	87.5	17.54386	100	100	10.44776	100	0	0	0	1
12	C11	100	36.38095	15.76786	100	100	61.53846	87.5	0.877193	100	100	14.92537	100	0	16	0	1
13	C12	100	23.01587	3.125	50	0	69.23077	87.5	3.368421	100	100	4.477612	100	0	19.8	0	1
14	C13	0	15.07937	20.98214	50	0	76.92308	87.5	11.12281	100	100	14.92537	0	0	6	0.245	1
15	C14	100	12.8254	0.892857	50	0	23.07692	37.5	2.333333	100	0	0	100	0	0	0	1
16	C15	0	8.603175	30.66071	50	0	15.38462	37.5	2.631579	100	100	10.44776	0	0	4.8	0	1

Figure 11-15. *Normalize the attributes*

4. Inside cell A648, type "Variance".

5. Enter =VAR.P(A2:B647) in B648 and then autofill to P648. Part of our worksheet should look like Figure 11-16.

638	C637	0	10.04762	4.464286	50	0	7.692308	87.5	0	0	0	0	0	0	0	0
639	C638	100	9.126984	1.035714	50	0	61.53846	87.5	1.017544	0	0	0	0	0	14	0.364
640	C639	100	22.34921	3.571429	100	100	23.07692	37.5	10.52632	0	0	0	0	0	8.8	0.537
641	C640	100	35.98413	2.678571	100	100	23.07692	87.5	2.052632	0	0	0	0	0	12	0.003
642	C641	100	42.5873	11.75	50	0	69.23077	87.5	12.2807	0	0	0	100	100	20	0
643	C642	100	11.63492	36.01786	100	100	30.76923	37.5	4.385965	0	0	0	0	0	13	0
644	C643	0	14.15873	2.678571	50	0	7.692308	87.5	7.017544	0	100	2.985075	100	0	10	0.394
645	C644	0	18.25397	48.21429	100	100	38.46154	25	7.017544	0	100	1.492537	100	0	10	0.001
646	C645	100	6.619048	0.732143	50	0	0	87.5	0.140351	0	0	0	0	0	14	0.75
647	C646	100	33.73016	12.05357	50	0	7.692308	37.5	29.08772	0	0	0	100	0	0	0
648	Variance	2154.938	1662.148	368.4599	884.1481	1490.052	1547.457	1124.415	1328.234	1833.264	2498.652	1671.513	1718.914	1974.792	408.2822	64.79672
649																

Figure 11-16. *The variances of each normalized attributes*

6. From Figure 11-16, we can tell that all attributes have fair amount of variances except for A15. Clearly, A1 has the largest variance.

7. Our next step is to create a correlation matrix that shows the correlation between every two attributes. A high positive or negative correlation coefficient between two attributes indicates that these two attributes are speaking about the same data story. To create a correlation matrix, create a new worksheet and rename this worksheet "correlation".

8. Inside the worksheet correlation, enter integers 1–16 inside cells B1:Q1, respectively. A quick way is to autofill these numbers.

9. Autofill integers 1–16 inside cells A2:A17.

10. Enter the following formula in B2, autofill to Q2 and then autofill together to B17:Q17:
    =CORREL(INDEX(normalized!$B$2:$Q$647,$A2,0),INDEX(normal ized!$B$2:$Q$647,B$1,0))

11. Type "A1" in cell B18 and then autofill to P18. Our worksheet should look like Figure 11-17.

⊿	A	B	C	D	E	F	G	H	I	J	K	L	M	N	O	P	Q
1		1	2	3	4	5	6	7	8	9	10	11	12	13	14	15	16
2	1	1	0.719911	0.613493	0.495427	0.517703	0.496117	0.232286	0.655401	0.402537	0.590902	0.664707	0.822962	0.801138	0.499181	0.65348	0.826006
3	2	0.719911	1	0.689071	0.503038	0.346346	0.24677	-0.21228	0.316199	0.452713	0.748915	0.446473	0.538864	0.869343	0.177529	0.817308	0.544116
4	3	0.613493	0.689071	1	0.740839	0.55716	0.543265	-0.08431	0.533774	0.686241	0.555444	0.384094	0.452082	0.714819	0.423602	0.538371	0.346217
5	4	0.495427	0.503038	0.740839	1	0.408331	0.458376	-0.01124	0.576071	0.704719	0.551195	0.382477	0.406827	0.67844	0.410414	0.620485	0.578942
6	5	0.517703	0.346346	0.55716	0.408331	1	0.799385	0.265171	0.719715	0.302346	0.124766	0.651829	0.349997	0.255741	0.508366	0.216649	0.383316
7	6	0.496117	0.24677	0.543265	0.458376	0.799385	1	0.594868	0.573061	0.683219	0.448457	0.822218	0.617129	0.295043	0.753858	0.14975	0.318477
8	7	0.232286	-0.21228	-0.08431	-0.01124	0.265171	0.594868	1	0.417614	0.372017	0.207738	0.538699	0.59275	-0.15633	0.770127	-0.17752	0.248532
9	8	0.655401	0.316199	0.533774	0.576071	0.719715	0.573061	0.417614	1	0.362152	0.196258	0.503612	0.564603	0.323996	0.749068	0.379335	0.717586
10	9	0.402537	0.452713	0.686241	0.704719	0.302346	0.683219	0.372017	0.362152	1	0.810422	0.5397	0.639834	0.530964	0.671582	0.369896	0.269994
11	10	0.590902	0.748915	0.555444	0.551195	0.124766	0.448457	0.207738	0.196258	0.810422	1	0.616512	0.768437	0.787479	0.468437	0.690523	0.478955
12	11	0.664707	0.446473	0.384094	0.382477	0.651829	0.822218	0.538699	0.503612	0.5397	0.616512	1	0.809627	0.514908	0.678309	0.504225	0.606532
13	12	0.822962	0.538864	0.452082	0.406827	0.349997	0.617129	0.59275	0.564603	0.639834	0.768437	0.809627	1	0.63579	0.777462	0.558675	0.71836
14	13	0.801138	0.869343	0.714819	0.67844	0.255741	0.295043	-0.15633	0.323996	0.530964	0.787479	0.514908	0.63579	1	0.213064	0.86732	0.677805
15	14	0.499181	0.177529	0.423602	0.410414	0.508366	0.753858	0.770127	0.749068	0.671582	0.468437	0.678309	0.777462	0.213064	1	0.238574	0.488993
16	15	0.65348	0.817308	0.538371	0.620485	0.216649	0.14975	-0.17752	0.379335	0.369896	0.690523	0.504225	0.558675	0.86732	0.238574	1	0.748781
17	16	0.826006	0.544116	0.346217	0.578942	0.383316	0.318477	0.248532	0.717586	0.269994	0.478955	0.606532	0.71836	0.677805	0.488993	0.748781	1
18		A1	A2	A3	A4	A5	A6	A7	A8	A9	A10	A11	A12	A13	A14	A15	

Figure 11-17. *The correlation matrix is computed*

12. The first thing we should look at is either column Q or row 17, which marks the correlation coefficients between the target and other attributes. We shall realize that A1 has the largest correlation coefficient with the target, indicating that A1 shares much data story with the target, that is, A1 could be a good indicator of the target. Since A1 also has the largest variance, certainly we should select A1 into the core feature set.

13. After A1 is selected, we should then look for independent attributes that share a small correlation coefficient with A1. A good candidate is A7. Let's take it.

14. Next candidate would be A9. For A9, we must examine its correlation coefficients with A1 and A7.

15. Should we take A4 or A14? Well, their correlation coefficients with A1 are on the mediate high, but their correlation coefficients with A9 or A7 are high (0.70 and 0.77, respectively). So, the core feature set could be just A1, A7, and A9.

The preceding operations are recorded in Chapter11-4b.xlsx. At this point, a good exercise is to try a data mining method with all the features or the core features only. Remember to apply cross-validation to assess if the core feature set can perform well.

Chapter11-5a.xlsx and Chapter11-6a.xlsx are designed for you to conduct a cross-validation on your chosen data mining method with all the features or the core features only. I applied logistic regression for this purpose because features highly correlated with the target can help regression models such as linear regression or logistical regression (remember that the feature A1 has high positive correlation with the target). My work can be found in Chapter11-5b.xlsx and Chapter11-6b.xlsx. Note, since the target has only two classes (0, 1), it is a perfect case to apply logistic regression. Of course, other data mining methods such as K-NN and Naïve Bayes would be good choices, too.

Reinforcement Exercises

There are three sets of reinforcement exercises.

1. Chapter11-HW-1a.xlsx is for EDA practice. Please use it to practice the following (refer to Chapter11-HW-1a-withAnswers.xlsx when needed):

   - Number of records

   - Number of attributes

   - Data type of each attribute

   - Sum, mean, median, and variances of each numerical attribute

   - Value range of each numerical attribute

   - Unique values of categorical target and attributes

   - Class distribution of the target

2. Chapter11-HW-2a.xlsx is for data cleaning practice. You can find the standard answer inside Chapter11-HW-2a-withAnswers.xlsx.

3. Chapter11-HW-3a.xlsx is for feature selection practice. Note, for this practice, there is no standard answer, I only make suggestions inside the file Chapter11-HW-3a-withSuggestions.xlsx.

Review Points

Please review the following topics and skills:

1. The initial investigations in EDA

2. Detection of missing data and anomalies

3. Data cleaning

4. Encoding including one hot encoding and ordinal encoding

5. Normalization, standardization, and normalization by median

6. Feature selection of data mining

7. Excel feature Remove Duplicates

8. Excel functions COUNT, COUNTA, MAX, MIN, AVERAGE, MEDIAN, IF, OR, ISNONTEXT, ISTEXT, ROW, CORREL, VAR.P, STDEV.P, and INDEX

CHAPTER 12

Association Analysis

We have learned that correlation measures the linear association strength between two numerical variables. While correlation study requires the two variables to be numeric and only measures the linear relationship, association analysis has no such restriction. Association analysis measures the strength of co-occurrence between two or more variables. It is an effective data mining technique for cross-selling, and it also has applications in biomedical research such as gene network study.

Please download the sample Excel files from `https://github.com/hhohho/Learn-Data-Mining-through-Excel-2` for this chapter's exercises.

General Understanding

The most well-known application of association analysis is the so-called market basket analysis, which finds the co-occurrence of one store item (or more) with another item within the same purchase transaction. By identifying what items are commonly purchased together, the stores (usually retail stores or supermarkets) can take relevant actions to increase their sales. Such actions include placing certain items together, promoting bundle prices, etc.

The common understanding is to find the co-occurrences among those frequently purchased items. Therefore, the first concept we need to understand is "support percentage", or simply speaking, "support". The support of an item (or itemset) is defined as the occurrence frequency of the item (or itemset) in the transaction set. I will use the data in Table 12-1 to help explain the concept of *support*.

Table 12-1. *A sample shopping basket*

Shopping basket	Beer	Movie	Diaper	Pen	Rice	Apple	Juice
1	1		1				
2	1	1		1		1	
3						1	1
4			1				
5	1	1					
6	1		1		1		
7							
8	1	1	1				
9	1	1					
10					1	1	1
11	1		1				
12		1					1

In Table 12-1, a shopping basket represents a purchase transaction and there are 12 transactions. Number 1 indicates that the item appears in the specific transaction. As Beer shows up 7 times out of the 12 transactions, support of Beer = 7/12. Similarly, support of Pen = 1/12, and support of Apple = 3/12.

Another important concept is "confidence". Confidence is similar to conditional probability. For example, when Beer is purchased, how often Diaper is also purchased? This probability is expressed as confidence (B → D) (Beer implies Diaper). Since when Beer is purchased, Diaper is also purchased 4 times; thus, confidence (B → D) = 4/7.

Note that confidence (D → B) = 4/5, as Diaper is purchased 5 times and there are 4 times Beer is purchased together with Diaper. This shows that itemset (B, D) is different from itemset (D, B).

Follow these instructions to experience how association analysis works:

1. Assume that the minimum support is 0.25 and the minimum confidence is 0.6.

2. Since Pen and Rice both have support values lower than 0.25, any itemsets containing either Pen or Rice would have a support value less than 0.25, too. Our interested items are then limited to Beer, Movie, Diaper, Apple, and Juice. Thus, the data to analyze are reduced as shown in Table 12-2.

Table 12-2. *Shopping basket data reduced after applying minimum support*

Shopping basket	Beer	Movie	Diaper	Apple	Juice
1	1		1		
2	1	1		1	
3				1	1
4			1		
5	1	1			
6	1		1		
7					
8	1	1	1		
9	1	1			
10				1	1
11	1		1		
12		1			1

3. Assume that we are only interested in itemsets containing at least two items. There are totally 20 different itemsets of size 2 $\left(\binom{5}{2} \times 2 \right)$.

Assessing them by hand would be very tedious and error prone, though it is still doable. Only four itemsets (Movie, Beer), (Diaper, Beer), (Apple, Juice), and (Juice, Apple) pass the confidence requirement. However, both (Apple, Juice) and (Juice, Apple) fail the support criteria as their support values are less than 0.25, which leaves only two itemsets of size 2. The result is illustrated in Table 12-3.

Table 12-3. *Itemsets of size 2 only*

Itemset	Support	Confidence	Itemset	Support	Confidence
(B, M)	4/10	4/7	(M, A)	1/10	1/5
(M, B)	4/10	4/5	(A, M)	1/10	1/3
(B, D)	4/10	4/7	(M, J)	1/10	1/5
(D, B)	4/10	4/5	(J, M)	1/10	1/3
(B, A)	1/10	1/7	(D, A)	0	0
(A, B)	1/10	1/3	(A, D)	0	0
(B, J)	0	0	(D, J)	0	0
(J, B)	0	0	(J, D)	0	0
(M, D)	1/10	1/5	(A, J)	2/10	2/3
(D, M)	1/10	1/5	(J, A)	2/10	2/3

4. Examining itemsets of size 3 can now only happen inside the three items: Beer, Movie, and Diaper. There are six different itemsets: (B, {M, D}), (M, {B, D}), (D, {B, M}) and ({M, D}, B), ({B, D}, M), ({B, M}, D). However, Beer, Movie, and Diaper co-occur only once; itemsets of size 3 would all fail the minimum support requirement, that is, there is no itemset of size 3 that meets the requirements. When there is no itemset of size 3, definitely there is no itemset of size 4 or 5 that can pass the criteria.

Conclusion: The preceding association analysis identified only two meaningful itemsets, (Movie, Beer) and (Diaper, Beer), based on the given minimum support and confidence values.

Learn Association Analysis Through Excel

Let's open Chapter12-1a.xlsx which contains 1831 records for 371 receipts (transactions). The 1831 records were selected and preprocessed from a much larger online retail dataset which is provided by UCI Machine Learning Repository at https://doi.org/10.24432/C5BW33. Part of the data looks like Figure 12-1.

	A	B	C	D	E
1	ReceiptNo	Quantity	ItemCode		
2	1	2	I28		
3	1	6	I46		
4	1	6	I47		
5	1	8	I48		
6	1	6	I50		
7	2	6	I20		
8	2	6	I21		
9	3	6	I26		
10	3	6	I27		
11	3	32	I44		
12	4	6	I38		

Figure 12-1. *Market basket data*

The data in Chapter12-1a.xlsx are sorted based on Receipt numbers. Column A presents the receipt numbers, column C shows the items on a receipt and there are 50 distinct items numbered from I1 to I50, and column B lists the purchased quantities of each item on a receipt. For example, rows 2–6 showcase items I28, I46, I47, I48, and I50 are on Receipt 1. Let's ignore the purchase quantity in this study.

For association analysis, the data in Chapter12-1a.xlsx must be reorganized as Table 12-1. Follow these instructions to organize the data in Chapter12-1a.xlsx:

1. Enter "Receipt#" in cell D1.

2. As there are 371 receipts numbering from 1 to 371, enter 1 in cell D2 and 2 in cell D3. Select both cells D2 and D3; autofill down to cell D372.

3. Enter "I1" in cell E1, then horizontally autofill from cell E1 to cell BB1. The cells E1:BB1 contain the item numbers I1, I1, ..., I50. Part of our worksheet looks like Figure 12-2.

⊿	A	B	C	D	E	F	G	H	I	J	K	L
1	ReceiptNo	Quantity	ItemCode	Receipt#	I1	I2	I3	I4	I5	I6	I7	I8
2	1	2	I28	1								
3	1	6	I46	2								
4	1	6	I47	3								
5	1	8	I48	4								
6	1	6	I50	5								
7	2	6	I20	6								
8	2	6	I21	7								
9	3	6	I26	8								
10	3	6	I27	9								
11	3	32	I44	10								
12	4	6	I38	11								
13	5	12	I24	12								
14	5	24	I25	13								
15	6	80	I8	14								
16	7	6	I20	15								

Figure 12-2. *Organize the data*

4. An item showing up on a receipt should have numerical value 1 in the corresponding cell. For example, cell L7 should have value 1 as item I8 is on receipt #6. Similarly, since I28 is on receipt #1, cell AF2 should have value 1, too. Cells E2, F2, and G2 should have value 0 inside as items I1, I2, and I3 are not on receipt #1. We need a formula to assign 1 under the item if it is on a receipt; otherwise, assign 0. Enter in cell E2 the following formula:

   =COUNTIFS($A$2:$A$1832,$D2,$C$2:$C$1832,E$1)

   In this formula, "$A$2:$A$1832,$D2" returns A2:A6 because $D2 matches 1. However, none of the cells inside A2:A6 has a value in column C equaling E$1 (I1). Thus, this formula returns 0 in cell E2.

5. Autofill the formula from E2 to cell BB2, and then autofill together to cells E372:BB372. Part of our data would look like Figure 12-3.

	C	D	E	F	G	H	I	J	K	L	M	N
1	ItemCode	Receipt#	I1	I2	I3	I4	I5	I6	I7	I8	I9	I10
2	I28	1	0	0	0	0	0	0	0	0	0	0
3	I46	2	0	0	0	0	0	0	0	0	0	0
4	I47	3	0	0	0	0	0	0	0	0	0	0
5	I48	4	0	0	0	0	0	0	0	0	0	0
6	I50	5	0	0	0	0	0	0	0	0	0	0
7	I20	6	0	0	0	0	0	0	0	1	0	0
8	I21	7	0	0	0	0	0	0	0	0	0	0
9	I26	8	0	0	0	0	0	0	0	0	0	0
10	I27	9	0	0	0	0	0	0	0	0	0	0
11	I44	10	0	0	0	0	0	0	0	0	0	0
12	I38	11	0	0	0	0	0	0	0	0	0	0
13	I24	12	0	1	0	0	0	0	0	0	0	0
14	I25	13	0	0	0	0	0	0	0	0	0	0
15	I8	14	0	0	0	0	0	0	0	1	0	0
16	I20	15	0	0	0	0	0	0	0	0	0	0
17	I21	16	0	0	0	0	0	0	0	0	0	0
18	I28	17	0	0	0	0	0	0	0	0	0	0
19	I41	18	0	0	0	0	0	0	0	0	0	0
20	I42	19	0	0	0	0	0	0	0	0	0	0
21	I45	20	0	0	0	0	0	0	0	0	0	0
22	I46	21	0	0	0	0	0	1	0	0	0	0

Figure 12-3. *Create the data table for association analysis*

We are going to examine all possible itemsets of size 2 in this example; therefore, we need to set up a table such that both the rows and columns of the table are marked by each item. Later, we also need to compute both the confidence and support values for each itemset. For that, we need to calculate (1) the number of times two items co-occur on a single receipt and (2) the number of receipts on which an item appears.

6. In cell BF1, enter the text "I1" and then autofill horizontally to cell DC1.

7. Enter the text "Occurrences" in BE2, enter the formula =COUNTIFS(E2:E372,1) in cell BF2, and autofill from cell BF2 to cell DC2. Cells BF2:DC2 reference "the number of receipts that an item appears on". For example, item I1 appears on 20 receipts (shown in Figure 12-4).

217

8. Enter 1 in cell BD3 and 2 in cell BD4. Select both cells BD3 and BD4; autofill them to cell BD52. This action will fill in numbers 1, 2, 3, ..., 50 into cells BD3, BD4, BD5, ..., BD52. These numbers will be used to reference a column in the data table E2:BB372 as there are 50 columns in E2:BB372, one column for one item.

9. Enter "I1" in cell BE3; autofill from cell BE3 to cell BE52.

   Compare our worksheet with Figure 12-4. If our worksheet looks different from Figure 12-4, we need to (1) examine our formulas and (2) make sure that there are no empty spaces in the item names. For example, in cell E1 or cell BF1, if the item name is entered as "I1 ", then the calculation for I1 won't be correct.

⬦	BB	BC	BD	BE	BF	BG	BH	BI	BJ	BK	BL	BM	BN	BO
1	I50				I1	I2	I3	I4	I5	I6	I7	I8	I9	I10
2		1		Occurrences	20	28	36	26	23	26	25	52	34	30
3		0	1	I1										
4		0	2	I2										
5		0	3	I3										
6		0	4	I4										
7		0	5	I5										
8		0	6	I6										
9		1	7	I7										
10		1	8	I8										
11		0	9	I9										
12		0	10	I10										
13		0	11	I11										
14		0	12	I12										
15		0	13	I13										
16		0	14	I14										
17		0	15	I15										
18		0	16	I16										

Figure 12-4. *Set up the association table*

10. Let's calculate the number of times two items co-occur in a single receipt. Enter the following formula in cell BF3:

    `=COUNTIFS(INDEX($E$2:$BB$372,0,$BD3),1,E$2:E$372,1)`

    BD3 = 1, so the function `INDEX($E$2:$BB$372,0,$BD3)` fetches the first column from the table E2:BB372, which is the column for item I1, that is, E2:E372. Thus, the preceding formula becomes `=COUNTIFS(E2:E372,1,E$2:E$372,1)` in cell BF3. This formula counts how many rows meeting the two conditions at the same time: cells inside E2:E372 having value 1 and cells inside E$2:E$372 having value 1, too.

In this specific formula, it happens that the two conditions are the same because cell BF3 is for itemset (I1, I1). This may be confusing. Let's continue; I will explain it more soon.

11. Autofill from cell BF3 to cell DC3, then autofill together to BF52:DC52. Let's take a look at the formula inside cell BG3. It is

`=COUNTIFS(INDEX($E$2:$BB$372,0,$BD3), 1,F$2:F$372,1)`

`INDEX($E$2:$BB$372,0,$BD3)` will still give us E2:E372 for item I1. However, `F$2:F$372` is for item I2. Thus, this formula is now counting how many times item I1 and item I2 co-appear on a single receipt.

Another example is the formula inside cell BJ4. It is

`=COUNTIFS(INDEX($E$2:$BB$372,0,$BD4), 1,I$2:I$372,1)`

As BD4 = 2, this formula is in fact the same as

`=COUNTIFS(F$2:F$372,1,I$2:I$372,1)`

Clearly, cell BJ4 is counting how many times item I2 and item I5 co-appear on a single receipt.

By now, let's compare our worksheet with Figure 12-5.

◢	BD	BE	BF	BG	BH	BI	BJ	BK	BL	BM
1			I1	I2	I3	I4	I5	I6	I7	I8
2		Occurrences	20	28	36	26	23	26	25	52
3	1	I1	20	1	2	3	1	3	2	4
4	2	I2	1	28	9	5	8	5	7	10
5	3	I3	2	9	36	12	9	11	10	13
6	4	I4	3	5	12	26	6	8	6	6
7	5	I5	1	8	9	6	23	8	8	9
8	6	I6	3	5	11	8	8	26	5	6
9	7	I7	2	7	10	6	8	5	25	9
10	8	I8	4	10	13	6	9	6	9	52
11	9	I9	3	6	15	13	10	13	8	8
12	10	I10	2	6	17	8	8	9	8	8
13	11	I11	4	10	12	10	9	9	7	12
14	12	I12	2	6	9	9	6	7	8	9
15	13	I13	4	9	8	8	5	5	6	8

Figure 12-5. *The co-occurrences of two items computed*

12. We need to construct another table to show what itemsets are valid based on our minimum support value and minimum confidence value.

- Enter the text "Support" in cell DF2, number 0.03 in cell DF3, text "Confidence" in cell DF4, and number 0.5 in cell DF5.

- Enter "I1" in cell DH1 and autofill horizontally to cell FE1.

- Enter "I1" in cell DG3 and autofill vertically to cell DG52.

Compare our table setup with Figure 12-6.

▲	DC	DD	DE	DF	DG	DH	DI	DJ	DK	DL	DM	DN
1	I50					I1	I2	I3	I4	I5	I6	I7
2	60			Support								
3	5			0.03	I1							
4	7			Confidence	I2							
5	9			0.5	I3							
6	7				I4							
7	7				I5							
8	7				I6							
9	8				I7							
10	10				I8							
11	8				I9							
12	10				I10							
13	12				I11							
14	8				I12							
15	8				I13							

Figure 12-6. *Association analysis table setup*

13. Enter the following formula in cell DH3, autofill from DH3 to FE3, and then autofill together from cells DH3:FE3 to cells DH52:FE52:

```
=IF($DG3=DH$1,"",IF(AND(BF$2>0,BF3/371>=$DF$3, BF3/BF$2>=$DF$5),
TEXT(BF3/371,"0.000") & ", " &TEXT(BF3/BF$2,"0.000"),""))
```

For this formula, we need to understand the following:

a. Note, cell DH3 references itemset (DH1, DG3) which happens to be (I1, I1). Similarly, cell DJ7 references itemset (DJ1, DG7) which is itemset (I3, I5).

b. If DG3 and DH1 are for the same item (they are in this case), there is no need to compute the support or confidence values.

c. BF$2>0 is to avoid the possible "divided by zero" error.

d. BF3/371>=$DF$3 is to make sure that the support value of the itemset (DH1, DG3) is no less than the minimum support value. Note that I hard-coded the number of receipts 371 here. A better approach is to use a cell reference, that is, storing the number of receipts in a cell. Later in the "Reinforcement Exercises," you will notice that I do store the number of receipts in a cell.

e. BF3/BF$2>=$DF$5 is to make sure that the confidence value of the itemset (DH1, DG3) meets the minimum confidence requirement.

f. The AND function guarantees if any requirement is not met, nothing is displayed in cell DH3.

g. TEXT function is used to format the results.

14. Compare our result with Figure 12-7. Adjust the minimum support value and confidence value to view the changes of found itemsets.

⊿	DF	DG	DH	DI	DJ	DK	DL	DM
1			I1	I2	I3	I4	I5	I6
2	Support							
3	0.03	I1						
4	Confidence	I2						
5	0.5	I3						
6		I4						
7		I5						
8		I6						
9		I7						
10		I8						
11		I9				0.035, 0.500		0.035, 0.500
12		I10						
13		I11						
14		I12						

Figure 12-7. *The association analysis of itemsets of size 2*

The final result is available in Chapter12-1b.xlsx.

At step 13, we can also enter a very complicated formula to generate the association analysis result as shown in Figure 12-7. Doing this way, we don't need to create the third table. Open the file Chapter12-2b.xlsx to view the complicated formula.

Using Excel to conduct association analysis on itemsets of size 3 or larger becomes impractical, that is why this chapter stops here. Please be aware that association analysis can be computationally expensive when the numbers of items and transactions are large.

Reinforcement Exercises

For this chapter, I have one exercise for you. Open the file Chapter12-HW.xlsx; you will see some data taken and preprocessed from the Online Retail dataset that can be downloaded at https://doi.org/10.24432/C5BW33 from UCI Machine Learning Repository. Take a look at Figure 12-8 which is a screenshot of the data.

	A	B	C	D	E	F
1	InvoiceNo	Quantity	StockCode			
2	1	12	s21			
3	2	2	s2			
4	2	1	s4			
5	2	2	s8			
6	2	3	s19			
7	2	2	s25			
8	2	2	s9			
9	3	2	s4			
10	3	2	s24			
11	3	1	s7			
12	3	2	s24			

Figure 12-8. *The data for the reinforcement exercise*

Certainly, the data inside Chapter12-HW.xlsx are organized the same as those in Chapter12-1a.xlsx except for the names of the data. Remember to treat Invoices as Receipts and Stock Code as Item Code. The file Chapter12-HW-withAnswers.xlsx contains my answer to this exercise. Be advised that inside the answer worksheet, the number of distinct invoices is stored in cell D2 instead of being hard-coded in formulas.

Hope you enjoy the exercise.

Review Points

1. Association and its potential applications

2. Item and itemset

3. Minimum support and minimum confidence

4. Functions IF, AND, COUNTIFS, INDEX, and TEXT and the operator &

## CHAPTER 13

Artificial Neural Network

A data mining book without artificial neural network is not complete. By its name, we can tell that artificial neural network mimics the architecture and biological process of human neurons. In our brain, millions of neurons work together to generate corresponding outputs from various inputs. This conceptual understanding is taken to develop artificial neural network method, though how neurons work together biologically still needs much research to elucidate.

Please download the sample Excel files from https://github.com/hhohho/Learn-Data-Mining-through-Excel-2 for this chapter's exercises.

General Understanding

Despite the extremely complicated biological architecture and process of our brain's neuron network, the neural network data mining method has clear mathematical foundation. In its simplest form, artificial neural network is a linear function, similar to linear discriminant analysis. It takes inputs in the form of attributes and searches for optimal coefficients (or weights) for the attributes to generate an output that is as close as possible to the experimental data. However, neural network can be much more complicated and flexible than linear discriminant analysis. Of course, it can also be nonlinear. The true beauty of neural network is that it mimics the learning ability of neurons, that is, the output can be fed back as the input to further optimize the weights.

The architecture in a typical neural network has multiple layers. The first layer is the input layer, and the last layer is the output layer. Between these two layers, there can be one or more hidden layers. The data into a layer are input to this specific layer; the information coming out of the layer is the output of the layer but the input to the next layer. At each hidden layer, data go through certain processing actions including aggregation and transformation. Figure 13-1 depicts a simple neural network architecture. In this architecture, all data go into every node.

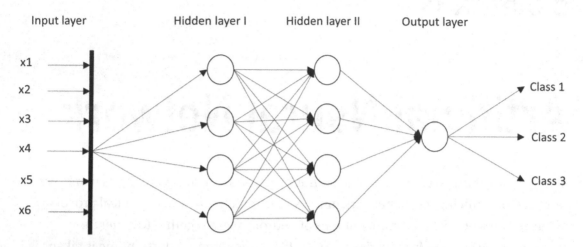

Figure 13-1. *Illustration of a simple neural network*

Hidden layers and the output layer are active layers because they aggregate and transform data. The transformation function is also called the activation function. The aggregate function can be linear or nonlinear. The two most commonly used activation functions are the sigmoid function and rectified linear unit function, though there are other activation functions such as bell-shaped function, logistic function, etc. Heaviside step function is also used commonly as it behaves very similarly to how a neuron generates output: either none or all. A common characteristic of these activation functions is that they all can provide a linear transformation, either for all data or only for a particular range of values. Clearly, these activation functions can simplify data by mapping the multidimensional data points into a linear space.

Equation (6-2) defines the sigmoid function in its simplest form. Let's review it again as Equation (13-1):

$$S = \frac{1}{1 + e^{-(m_1 x_1 + m_2 x_2 + \cdots + m_n x_n + b)}} \tag{13-1}$$

Figure 13-2 shows the graph of the sigmoid function. Note that sigmoid function converges to 1 or -1 when the input goes to the positive or negative infinity. More importantly, sigmoid function can transform data into a linear space if the input is between -1 and 1. That is why data input of sigmoid activation function is usually normalized to the range of -1 and 1, or at least, the range of 0 and 1.

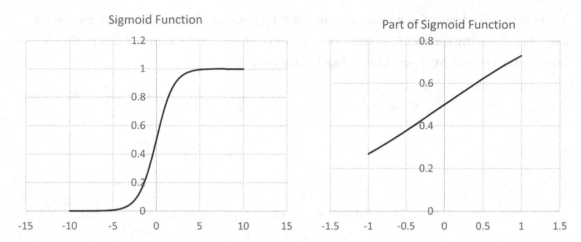

Figure 13-2. *Sigmoid function is linear between -1 and 1*

Learn Neural Network Through Excel

We are going to practice two experiments in this chapter. The first experiment is much simpler. It can help us to get a good idea about how a neural network works. The second experiment reinforces and elevates our understanding.

Experiment 1

Open Chapter13-1a.xlsx which contains the Breast Cancer Wisconsin (Original) dataset downloaded from UCI Machine Learning Repository at https://doi.org/10.24432/C5HP4Z. For details of the dataset, please visit https://archive.ics.uci.edu/ or read the journal article (W.H. Wolberg and O.L. Mangasarian: "Multisurface method of pattern separation for medical diagnosis applied to breast cytology," PNAS, 1990, 87:9193-9196).

There are originally ten attributes and one target named "Class". The class has two values: 2 for benign breast cancer and 4 for malignant breast cancer. The first attribute is the sample code number which is removed in Chapter13-1a.xlsx. The other nine attributes are renamed x1, x2, ..., x9 for simplicity. The downloaded dataset had 699 samples originally. Because one attribute is missing in 16 samples, these 16 samples were removed. Thus in Chapter13-1a.xlsx, there are only 683 samples.

Five hundred samples are used as the training dataset; the rest 183 samples will be used as the testing dataset to assess how well our neural network model can perform. The top part of our worksheet looks like Figure 13-3.

◢	A	B	C	D	E	F	G	H	I	J	K	L	M	N
1	Neuron	w1	w2	w3	w4	w5	w6	w7	w8	w9	b	Wo	Bo	
2	1													
3	2													
4	3													
5														
6														
7														
8														
9												neuron weighted sum		
10		x1	x2	x3	x4	x5	x6	x7	x8	x9	Class	1	2	3
11		10	10	10	8	6	1	8	9	1	4			
12		1	1	1	1	2	1	3	1	1	2			
13		3	1	1	1	2	1	1	1	1	2			
14		3	1	1	1	2	1	2	1	1	2			
15		10	10	10	10	10	10	4	10	10	4			
16		6	5	5	8	4	10	3	4	1	4			
17		1	1	1	1	2	1	1	1	1	2			
18		5	2	1	1	2	1	1	1	1	2			
19		1	1	1	1	2	1	3	1	1	2			
20		4	10	8	5	4	1	10	1	1	4			
21		4	8	8	5	4	5	10	4	1	4			

Figure 13-3. *The breast cancer dataset downloaded from UCI*

In our first experiment, for the purpose of simplicity, we are going to have only one hidden layer with three neurons (usually called Nodes) as marked in cells A1:A4. The weights, w1, w2, ..., w9, are the weights (coefficients) for the nine attributes, while b is the so-called bias, that is, the intercept in a linear function. Each neuron has its own set of weights and bias. Note, we are going to use a linear aggregation function. Wo and Bo represent the weights and bias used for the output layer. If confused at this moment, don't worry. I will explain more details at each step when these data are used.

Follow these instructions to complete this experiment:

1. The first task is to assign values to the weights and intercepts. Let's assign 1s to all of them as shown in Figure 13-4.

⊿	A	B	C	D	E	F	G	H	I	J	K	L	M	N
1	Neuron	w1	w2	w3	w4	w5	w6	w7	w8	w9	b	Wo	Bo	
2	1	1	1	1	1	1	1	1	1	1	1	1	1	
3	2	1	1	1	1	1	1	1	1	1	1	1	1	
4	3	1	1	1	1	1	1	1	1	1	1	1	1	
5														
6														
7														
8														
9												neuron weighted sum		
10		x1	x2	x3	x4	x5	x6	x7	x8	x9	Class	1	2	3
11		10	10	10	8	6	1	8	9	1	4			
12		1	1	1	1	2	1	3	1	1	2			
13		3	1	1	1	2	1	1	1	1	2			
14		3	1	1	1	2	1	2	1	1	2			
15		10	10	10	10	10	10	4	10	10	4			
16		6	5	5	8	4	10	3	4	1	4			
17		1	1	1	1	2	1	1	1	1	2			
18		5	2	1	1	2	1	1	1	1	2			
19		1	1	1	1	2	1	3	1	1	2			
20		4	10	8	5	4	1	10	1	1	4			

Figure 13-4. *Initialize weights and intercepts with 1s*

2. Each neuron takes all training samples as input. The training samples are stored in table B11:K510. For sample 1 (row 11), all of its attribute values are aggregated into a single value inside neurons 1, 2, and 3, respectively. Thus, we need to enter the following formula into cell L11:

    =SUMPRODUCT($B11:$J11,INDEX($B$2:$J$4,L$10,0))+INDEX
    ($K$2:$K$4,L$10,1)

    This formula implements the linear expression w1x1 + w2x2 + w3x3 + ... + b. In this specific formula, w1, w2, w3... and b are for neuron 1 (B2:K2), and x1, x2, x3, ... are for the first data point in row 11.

    The function SUMPRODUCT is used to compute the sum product of the two arrays: B11:J11 and B2:J2. We can write the function as SUMPRODUCT($B11:$J11,$B$2:$J$2). However, we want to autofill from L11 to N11 without rewriting the formula again and again. Therefore, INDEX($B$2:$J$4,L$10,0) is applied to obtain the array B2:J2 based on the value in cell L10. As L10 = 1, INDEX($B$2:$J$4,L$10,0) fetches the first row from the table B2:J4.

INDEX($K$2:$K$4,L$10,1) fetches the value inside cell K2, that is, the bias. Note, there is only one column in the array K2:K4; that is why the last parameter inside the INDEX function is 1.

3. Autofill from L11 to N11, and then autofill together to L510:N510. Part of our worksheet looks like Figure 13-5. Examine the formula inside cell M11. Since M10 = 2, the aggregation function makes use of the coefficients of neuron 2.

    This step can be understood as the input layer.

◢	E	F	G	H	I	J	K	L	M	N
1	w4	w5	w6	w7	w8	w9	b	Wo	Bo	
2	1	1	1	1	1	1	1	1	1	
3	1	1	1	1	1	1	1	1	1	
4	1	1	1	1	1	1	1	1	1	
5										
6										
7										
8										
9									neuron weighted sum	
10	x4	x5	x6	x7	x8	x9	Class	1	2	3
11	8	6	1	8	9	1	4	64	64	64
12	1	2	1	3	1	1	2	13	13	13
13	1	2	1	1	1	1	2	13	13	13
14	1	2	1	2	1	1	2	14	14	14
15	10	10	10	4	10	10	4	85	85	85
16	8	4	10	3	4	1	4	47	47	47
17	1	2	1	1	1	1	2	11	11	11

Figure 13-5. *Aggregation function applied*

As mentioned before, after the aggregation function, the activation function (sigmoid function) needs to be applied to start the hidden layer. However, since sigmoid function favors its input data to be normalized between -1 and 1 or between 0 and 1, we need to normalize the values in columns L, M, and N. Let's just normalize the values into the range of 0 and 1 as all values are positive at this point.

4. Enter "MAX" inside cell K6 and "MIN" in cell K7.

5. Enter the formula =MAX(L11:L510) in cell L6 and the formula
    =MIN(L11:L510) in cell L7.

6. Select both cells L6 and L7; autofill to N6:N7 as shown in
    Figure 13-6.

7. In addition, merge cells O9:Q9; enter the text "normalized"
    in the merged cell, and 1, 2, and 3 in cells O10, P10, and Q10,
    respectively. The normalized weighted sums will be stored in
    columns O, P, and Q, respectively.

    Compare our worksheet with Figure 13-6. Don't be surprised that
    numbers in columns L, M, and N are the same.

	H	I	J	K	L	M	N	O	P	Q
1	w7	w8	w9	b	Wo	Bo				
2	1	1	1	1	1	1				
3	1	1	1	1	1	1				
4	1	1	1	1	1	1				
5										
6				MAX	85	85	85			
7				MIN	10	10	10			
8										
9					neuron weighted sum			normalized		
10	x7	x8	x9	Class	1	2	3	1	2	3
11	8	9	1	4	64	64	64			
12	3	1	1	2	13	13	13			
13	1	1	1	2	13	13	13			
14	2	1	1	2	14	14	14			
15	4	10	10	4	85	85	85			

Figure 13-6. Ready to normalize data

8. Enter the formula =(L11-L$7)/(L$6-L$7) in cell O11. This formula
    normalizes the value inside cell L11 to be between 0 and 1 inclusively.

9. Autofill from O11 to Q11, and then autofill together to O510:Q510.
    Don't be surprised that numbers in columns O, P, and Q are the
    same. Be aware that the formula =(L11-L$7)/(L$6-L$7) can
    cause an error when L6 (the maximum) equals L7 (the minimum).
    Such a case should never happen, however. If it does happen, that
    means the attribute has no variance and should be removed.

10. Time to transform the normalized data by the sigmoid function. Merge cells R9, S9, and T9. In the merged cell, enter "sigmoid transformed". In cells R10, S10, and T10, enter numbers 1, 2, and 3, respectively. Part of our worksheet looks like Figure 13-7.

◢	K	L	M	N	O	P	Q	R	S	T
1	b	Wo	Bo							
2		1	1	1						
3		1	1	1						
4		1	1	1						
5										
6	MAX		85	85	85					
7	MIN		10	10	10					
8										
9			neuron weighted sum			normalized			sigmoid transformed	
10	Class		1	2	3	1	2	3	1	2	3
11		4	64	64	64	0.72	0.72	0.72			
12		2	13	13	13	0.04	0.04	0.04			
13		2	13	13	13	0.04	0.04	0.04			
14		2	14	14	14	0.05333	0.05333	0.05333			

Figure 13-7. *After normalization but before sigmoid transformation*

11. Based on Equation (11-1), enter the formula =1/(1+EXP(-O11)) in cell R11.

12. Autofill from R11 to T11, then autofill together to R510:T510.

13. In cells U10 and V10, enter "Output" and "Error", respectively. By now, part of our worksheet looks exactly like Figure 13-8.

	L	M	N	O	P	Q	R	S	T	U	V
1	Wo	Bo									
2	1	1									
3	1	1									
4	1	1									
5											
6	85	85	85								
7	10	10	10								
8											
9	neuron weighted sum			normalized			sigmoid transformed				
10	1	2	3	1	2	3	1	2	3	Output	Error
11	64	64	64	0.72	0.72	0.72	0.67261	0.67261	0.67261		
12	13	13	13	0.04	0.04	0.04	0.51	0.51	0.51		
13	13	13	13	0.04	0.04	0.04	0.51	0.51	0.51		
14	14	14	14	0.05333	0.05333	0.05333	0.51333	0.51333	0.51333		
15	85	85	85	1	1	1	0.73106	0.73106	0.73106		

Figure 13-8. *Ready to generate output and compute errors*

14. The output is another aggregation function using the Wo values in cells L2:L4 and the Bo value in cell M2. In cell U11, enter the following formula:

    =MMULT(R11:T11,$L$2:$L$4)+$M$2

    Note that the function SUMPRODUCT does not work for the two arrays R11:T11 and L2:L4. This is because R11:T11 is like a row, while L2:L4 is like a column. However, the function MMULT (matrix multiplication) is perfect for this type of array multiplication. Observe cells M3 and M4 are not used.

15. Autofill from cell U11 to cell U510.

16. In cell V11, enter the formula =(U11-K11)^2 which gives the square of the difference between the output U11 and the original Class value K11. We have seen the use of square in error calculation before. One reason to square the difference is to make the error positive.

17. Autofill from cell V11 to cell V510.

   In neural network, we need to search for optimal values for weights w1, w2, ..., w9, Wo, b, and Bo such that the sum of the errors is minimized.

18. In cell P1, enter "Error Sum".

19. In cell Q1, enter the formula =SUM(V11:V510).

   Compare our result with Figure 13-9.

◢	L	M	N	O	P	Q	R	S	T	U	V
1	Wo	Bo			Error Sum	314.828					
2	1	1									
3	1	1									
4	1	1									
5											
6	85	85	85								
7	10	10	10								
8											
9	neuron weighted sum			normalized			sigmoid transformed				
10	1	2	3	1	2	3	1	2	3	Output	Error
11	64	64	64	0.72	0.72	0.72	0.67261	0.67261	0.67261	3.01782	0.96468
12	13	13	13	0.04	0.04	0.04	0.51	0.51	0.51	2.53	0.2809
13	13	13	13	0.04	0.04	0.04	0.51	0.51	0.51	2.53	0.2809
14	14	14	14	0.05333	0.05333	0.05333	0.51333	0.51333	0.51333	2.53999	0.29159
15	85	85	85	1	1	1	0.73106	0.73106	0.73106	3.19318	0.65097

Figure 13-9. Output generated and error calculated

20. We need to make use of Solver again to find the optimal weights and biases. Click Data ➤ Solver; follow Figure 13-10 to set up the cell ranges for Solver.

21. Check "Make Unconstrained Variables Non-Negative". The result will be different (likely better) if we uncheck it. In this experiment, let's check it.

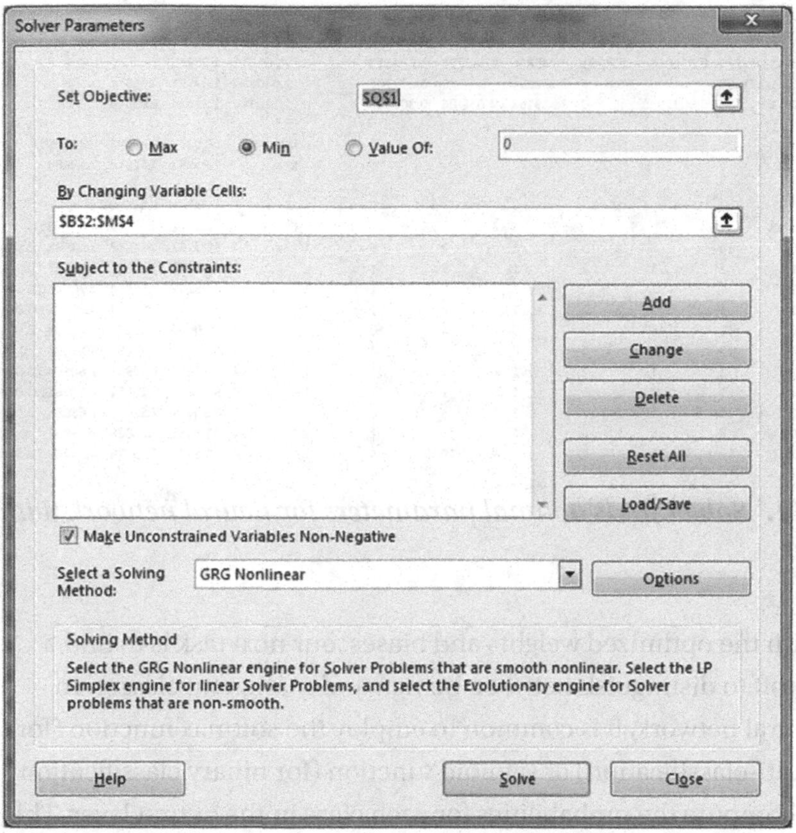

Figure 13-10. *Using Solver to find optimal parameters*

It may take some time for Solver to complete the optimal parameter search. Once Solver completes its work, our worksheet should look like Figure 13-11.

	B	C	D	E	F	G	H	I	J	K	L	M	N	O	P	Q
1	w1	w2	w3	w4	w5	w6	w7	w8	w9	b	Wo	Bo			Error Sum	199.853
2	0.015877	1.595424	1.547291	0.032066	0.06231	3.245841	0.032293	0.284619	0	1.04311	1.58285	0				
3	0.019502	1.648973	1.271313	0	3.64E-07	3.256054	0	0.292792	0	1.040896	1.64971	0.96571				
4	0.009889	1.406611	1.515587	0.002798	0	3.335183	4.03E-06	0.300405	0	1.116166	1.71624	1.09346				
5																
6										MAX	69.0122	65.9272	66.8209			
7										MIN	7.85883	7.52953	7.68664			
8																
9											neuron weighted sum			normalized		
10	x1	x2	x3	x4	x5	x6	x7	x8	x9	Class	1	2	3	1	2	3
11	10	10	10	8	6	1	8	9	1	4	39.3252	36.33	36.4983	0.51455	0.49318	0.48722
12	1	1	1	1	2	1	3	1	1	2	7.98573	7.52953	7.68665	0.00208	6.2E-09	1.4E-07
13	3	1	1	1	2	1	1	1	1	2	7.95289	7.56854	7.70642	0.00154	0.00067	0.00033
14	3	1	1	1	2	1	2	1	1	2	7.98519	7.56854	7.70642	0.00207	0.00067	0.00033
15	10	10	10	10	10	10	4	10	10	4	69.0066	65.9272	66.8209	0.99991	1	1
16	6	5	5	8	4	10	3	4	1	4	51.0515	49.491	50.3623	0.7063	0.71855	0.72167
17	1	1	1	1	2	1	1	1	1	2	7.92114	7.52953	7.68664	0.00102	6.2E-09	0
18	5	2	1	1	2	1	1	1	1	2	9.58007	9.25651	9.13281	0.02815	0.02957	0.02446
19	1	1	1	1	2	1	3	1	1	2	7.98573	7.52953	7.68665	0.00208	6.2E-09	1.4E-07
20	4	10	8	5	4	1	10	1	1	4	33.7021	31.328	30.9961	0.4226	0.40752	0.39418
21	4	8	8	5	4	5	10	4	1	4	44.3485	41.9326	42.4249	0.59669	0.58912	0.58745

Figure 13-11. *Solver finds optimal parameters for neural network with one hidden layer*

With the optimized weights and biases, our next task is to find a cutoff to distinguish between benign and malignant classes. In neural network, it is common to employ the softmax function (for multi-classification) or sigmoid function (for binary classification) to compute the probabilities for each class in the output layer. This requires quite some effort for the computation of cross-entropy. For simplicity, let's just find a cutoff to complete the output layer in our experiment. Follow these instructions to compute the cutoff:

22. Enter "mean" and "count" in cells P6 and P7, respectively.

23. Enter numbers 2 and 4 in cells Q5 and R5 and "cutoff" and "missed" in cells S5 and T5.

24. Enter the formula =AVERAGEIFS($U11:$U510,$K11:$K510,Q5) in cell Q6. This formula computes the averaged output value for class benign.

25. Autofill from Q6 to R6.

26. Enter the formula =COUNTIFS($K11:$K510,Q5) in cell Q7. This formula computes the number of class benign in the training dataset.

27. Autofill from Q7 to R7.

28. Enter the formula =(Q6*Q7+R6*R7)/500 in cell S6. This is the cutoff to distinguish class 2 and 4 (benign and malignant).

    Part of our worksheet looks like Figure 13-12.

◢	O	P	Q	R	S	T	U	V
1		Error Sum	199.853					
2								
3								
4								
5			2	4	cutoff	missed		
6		mean	2.52474	3.27248	2.79841			
7		count	317	183				
8								
9		normalized			sigmoid transformed			
10	1	2	3	1	2	3	Output	Error
11	0.51455	0.49317747	0.48722	0.62587	0.62085	0.61945	3.07802	0.85005
12	0.00208	6.23489E-09	1.4E-07	0.50052	0.5	0.5	2.47522	0.22583
13	0.00154	0.000667909	0.00033	0.50038	0.50017	0.50008	2.47542	0.22603
14	0.00207	0.000667909	0.00033	0.50052	0.50017	0.50008	2.47563	0.22623

Figure 13-12. *Cutoff computed for the neural network model with one hidden layer*

It's time to assess our neural network model (note again that this model is just a set of optimized parameters) with the testing dataset. Follow these instructions:

29. Select cells L510:U510; autofill them together down to cells L693:U693.

30. Enter "Predict" and "Diff" in cells W10 and X10.

31. Enter the formula =IF(U11<=$S$6,2,4) in cell W11; autofill from W11 to W693.

32. Enter the formula =IF(K11=W11,0,1) in cell X11; autofill from X11 to X693.

33. Enter the formula =SUM(X511:X693) in cell T6. The number in
T6 tells how many testing samples are not correctly predicted by
our neural network model. Be careful that the cell range in this
formula is X511:X693.

Part of our worksheet looks exactly like Figure 13-13. There are
5 out of 183 testing samples that are not correctly predicted,
which accounts to the accuracy rate of 97.3%. You can find the
completed result in Chapter13-1b..xlsx.

▲	O	P	Q	R	S	T	U	V	W	X
1		Error Sum	199.853							
2										
3										
4										
5			2	4	cutoff	missed				
6		mean	2.52474	3.27248	2.79841	5				
7		count	317	183						
8										
9		normalized			sigmoid transformed					
10	1	2	3	1	2	3	Output	Error	Predict	Diff
11	0.51455	0.49317747	0.48722	0.62587	0.62085	0.61945	3.07802	0.85005	4	0
12	0.00208	6.23489E-09	1.4E-07	0.50052	0.5	0.5	2.47522	0.22583	2	0
13	0.00154	0.000667909	0.00033	0.50038	0.50017	0.50008	2.47542	0.22603	2	0
14	0.00207	0.000667909	0.00033	0.50052	0.50017	0.50008	2.47563	0.22623	2	0
15	0.99991	1	1	0.73104	0.73106	0.73106	3.61783	0.14605	4	0
16	0.7063	0.718547267	0.72167	0.66958	0.67229	0.67298	3.32391	0.45709	4	0
17	0.00102	6.23489E-09	0	0.50025	0.5	0.5	2.4748	0.22544	2	0

Figure 13-13. *Assess the neural network model with the testing dataset*

For this specific breast cancer dataset, both benign and malignant classes are well
characterized by the nine attributes. It is not difficult to differentiate between the two
classes. The purpose of the preceding step-by-step instructions in Excel is to dissect
the neural network modeling process and present us a clear image about how a simple
neural network model works.

As shown in Q7:R7, the target class distribution is not very well balanced. Since we
didn't balance the target class distribution, it is a good idea to examine how well our
model can accurately predict the underrepresented class (which is 4). A little bit more
analysis will soon confirm us that the accuracy rate is 94.6% for our model to predict the
underrepresented class, a fairly good performance.

Experiment 2

Let's start another experiment during which we are going to build a neural network with two hidden layers. Still, each hidden layer has three neurons. When there is more than one hidden layer, the neural network is usually called deep learning. Note, true deep learning usually has more than two hidden layers and much more neurons.

Open the file Chapter13-2a.xlsx. The data inside are experimental results for short hairpin RNA (shRNA) efficacy prediction. The data were preprocessed and simplified. There are four attributes used to predict the target Class which has three distinct values: 39 = low efficacy, 69 = medium efficacy, and 84 = high efficacy. Rows 11–182 are the training dataset, while rows 183–190 are the scoring dataset. The worksheet is set up like Figure 13-14. Note that different neurons start with different weights and biases. W21, W22, W23, and B2 are the weights and bias for the neurons in hidden layer 2. We will call the two hidden layers "layer 1" and "layer 2", respectively, hereafter.

▲	A	B	C	D	E	F	G	H	I	J	K	L	M	N	O
1	Neuron	w1	w2	w3	w4	b1	Wo1	Bo1	W21	W22	W23	B2	Wo2	Bo2	
2	1	1	1	1	1	1	1	1	1	1	1	1	1	1	
3	2	1.5	1.5	1.5	1.5	1.5	1.5	1.5	1.5	1.5	1.5	1.5	1.5	1.5	
4	3	2	2	2	2	2	2	2	2	2	2	2	2	2	
5															
6						max									
7						min									
8															
9							neuron weighted sum			normalized weighted sum			sigmoid transformed		
10		x1	x2	x3	x4	Class	1	2	3	1	2	3	1	2	3
11		30.1	-4.9	-0.4	8	39									
12		31.1	-5.6	-0.4	7	39									
13		44.4	-5.1	-3.7	5	39									
14		48.4	-4.5	-3.3	5	39									
15		33.1	-4.4	-1.8	7	39									
16		39.6	-6.9	-4.3	5	39									
17		44.4	-6.9	-3.8	5	39									
18		34.1	-4.4	-0.6	7	39									
19		47.4	-5.7	-4.6	5	39									
20		33.1	-4.5	-2.4	8	39									
21		34.1	-4	-2.5	8	39									
22		36.9	-5.3	-2.5	7	39									

Figure 13-14. *Experiment 2 data setup*

1. Enter in cell G11 the following formula:

   =SUMPRODUCT($B11:$E11,INDEX($B$2:$E$4,G$10,
   0))+INDEX($F$2:$F$4,G$10,1)

   Again, this formula implements the linear expression w1x1 + w2x2 + w3x3 + … + b.

2. Autofill from G11 to I11 and then autofill together to G182:I182.

3. To normalize the data into range [-1, 1], we need to normalize the data into range [0, 1] first, multiply by 2, and then subtract by 1. Thus, we need to do the following:

   • In cell G6, enter the formula =MAX(G11:G182); autofill from G6 to I6.

   • In cell G7, enter the formula =MIN(G11:G182); autofill from G7 to I7.

   • In cell J11, enter the formula =(G11-G$7)/(G$6-G$7)*2-1. Autofill from cell J11 to cell L182.

   Part of our worksheet looks like Figure 13-15.

	A	B	C	D	E	F	G	H	I	J	K	L
1	Neuron	w1	w2	w3	w4	b1	Wo1	Bo1	W21	W22	W23	B2
2	1	1	1	1	1	1	1	1	1	1	1	1
3	2	1.5	1.5	1.5	1.5	1.5	1.5	1.5	1.5	1.5	1.5	1.5
4	3	2	2	2	2	2	2	2	2	2	2	2
5												
6						max	60.6	90.9	121.2			
7						min	26	39	52			
8												
9							.	neuron weighted sum		normalized weighted sum		
10		x1	x2	x3	x4	Class	1	2	3	1	2	3
11		30.1	-4.9	-0.4	8	39	33.8	50.7	67.6	-0.54913	-0.54913	-0.54913
12		31.1	-5.6	-0.4	7	39	33.1	49.65	66.2	-0.5896	-0.5896	-0.5896
13		44.4	-5.1	-3.7	5	39	41.6	62.4	83.2	-0.09827	-0.09827	-0.09827
14		48.4	-4.5	-3.3	5	39	46.6	69.9	93.2	0.190751	0.190751	0.190751
15		33.1	-4.4	-1.8	7	39	34.9	52.35	69.8	-0.48555	-0.48555	-0.48555
16		39.6	-6.9	-4.3	5	39	34.4	51.6	68.8	-0.51445	-0.51445	-0.51445
17		44.4	-6.9	-3.8	5	39	39.7	59.55	79.4	-0.20809	-0.20809	-0.20809
18		34.1	-4.4	-0.6	7	39	37.1	55.65	74.2	-0.35838	-0.35838	-0.35838
19		47.4	-5.7	-4.6	5	39	43.1	64.65	86.2	-0.01156	-0.01156	-0.01156
20		33.1	-4.5	-2.4	8	39	35.2	52.8	70.4	-0.46821	-0.46821	-0.46821
21		34.1	-4	-2.5	8	39	36.6	54.9	73.2	-0.38728	-0.38728	-0.38728

Figure 13-15. *Aggregated data normalized to [-1, 1]*

Follow these instructions to finish layer 1 data transformation and output generation:

4. Enter the text "Output-1" in cell P10 and text "Error-1" in cell Q10.

5. Enter the formula =1/(1+EXP(-J11)) in cell M11.

6. Autofill from cell M11 to cell O11, then autofill together from cells M11:O11 to cells M182:O182.

7. In cell P11, enter the formula =MMULT(M11:O11,$G$2:$G$4)+$H$2. This is the output from hidden layer 1.

8. In cell Q11, enter the formula =(P11-F11)^2. This is the error of layer 1 with the current parameter settings.

9. Autofill P11 and Q11 to P182 and Q182, respectively.

10. In cell P1, enter the text "Error Sum1".

11. In cell Q1, enter the formula =SUM(Q11:Q182).

Part of our worksheet should look like Figure 13-16.

	F	G	H	I	J	K	L	M	N	O	P	Q
1	b1	Wo1	Bo1	W21	W22	W23	B2	Wo2	Bo2		Error Sum1	778010.9
2	1	1	1	1	1	1	1	1	1			
3	1.5	1.5	1.5	1.5	1.5	1.5	1.5	1.5	1.5			
4	2	2	2	2	2	2	2	2	2			
5												
6	max	60.6	90.9	121.2								
7	min	26	39	52								
8												
9		neuron weighted sum			normalized weighted sum			sigmoid transformed				
10	Class	1	2	3	1	2	3	1	2	3	Output-1	Error-1
11	39	33.8	50.7	67.6	-0.54913	-0.54913	-0.54913	0.366066	0.366066	0.366066	2.64729518	1321.519
12	39	33.1	49.65	66.2	-0.5896	-0.5896	-0.5896	0.356728	0.356728	0.356728	2.60527465	1324.576
13	39	41.6	62.4	83.2	-0.09827	-0.09827	-0.09827	0.475453	0.475453	0.475453	3.13953974	1285.973
14	39	46.6	69.9	93.2	0.190751	0.190751	0.190751	0.547544	0.547544	0.547544	3.46394704	1262.811
15	39	34.9	52.35	69.8	-0.48555	-0.48555	-0.48555	0.380943	0.380943	0.380943	2.71424187	1316.656
16	39	34.4	51.6	68.8	-0.51445	-0.51445	-0.51445	0.374151	0.374151	0.374151	2.68367824	1318.875
17	39	39.7	59.55	79.4	-0.20809	-0.20809	-0.20809	0.448164	0.448164	0.448164	3.01673709	1294.795
18	39	37.1	55.65	74.2	-0.35838	-0.35838	-0.35838	0.411351	0.411351	0.411351	2.85108137	1306.744
19	39	43.1	64.65	86.2	-0.01156	-0.01156	-0.01156	0.49711	0.49711	0.49711	3.23699436	1278.993
20	39	35.2	52.8	70.4	-0.46821	-0.46821	-0.46821	0.38504	0.38504	0.38504	2.73268203	1315.318

Figure 13-16. *Layer 1 output and error calculated*

In a neural network, every layer is supposed to optimize its parameter set before it passes on its output to the next layer as input. This is why we must compute the output and error sum of layer 1. We need to use Solver to optimize layer 1 parameters by minimizing the error sum of layer 1.

241

12. Click Data ➤ Solver; set up the parameters of Solver as shown
     in Figure 13-17. Leave "Make Unconstrained Variables Non-
     Negative" unchecked. Click Solve.

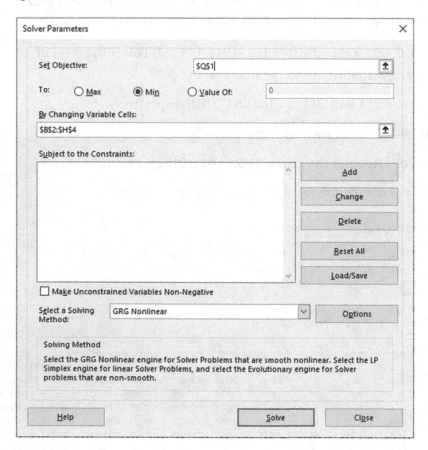

Figure 13-17. *Solver for hidden layer 1 parameter optimization*

13. On the upcoming menu, make sure the option "Keep Solver
     Solution" is chosen. By now, our result looks like Figure 13-18.

	A	B	C	D	E	F	G	H	I	J	K	L	M	N	O	R	Q
1	Neuron	w1	w2	w3	w4	b1	Wo1	Bo1	W21	W22	W23	B2	Wo2	Bo2			Error Sum1	26765
2	1	1.400606	-1.92613	2.120757	1.92234	1	13.4874	20.49571	1	1	1	1	1	1				
3	2	2.450927	-7.41431	4.679796	4.12889	1.5	29.64371	1.5	1.5	1.5	1.5	1.5	1.5	1.5				
4	3	4.316527	-16.2675	8.108013	7.05496	2	52.09065	2	2	2	2	2	2	2				
5																		
6						max	110.8898	229.4406	427.4011									
7						min	54.01284	116.3784	221.8142									
8																		
9							neuron weighted sum			normalized weighted sum			sigmoid transformed					
10		x1	x2	x3	x4	Class	1	2	3	1	2	3	1	2	3	Output-1	Error-1
11		30.1	-4.9	-0.4	8	39	67.12667	142.7622	264.8347	-0.53887	-0.53329	-0.58149	0.36845	0.369751	0.358591	55.1051516	259.3759
12		31.1	-5.6	-0.4	7	39	67.95323	146.2742	273.4835	-0.50981	-0.47116	-0.49735	0.375239	0.384342	0.378164	56.6488278	311.4811
13		44.4	-5.1	-3.7	5	39	74.77505	151.4633	281.8932	-0.26993	-0.37937	-0.41554	0.432925	0.406279	0.397585	59.088838	403.5614
14		48.4	-4.5	-3.3	5	39	80.07009	158.6904	292.642	-0.08373	-0.25153	-0.31097	0.479079	0.437448	0.422878	61.9528035	526.8312
15		33.1	-4.4	-1.8	7	39	65.47403	135.7272	251.2443	-0.59698	-0.65773	-0.7137	0.355034	0.341249	0.328783	52.5265983	182.9689
16		39.6	-6.9	-4.3	5	39	70.24672	150.2367	285.5905	-0.42916	-0.40107	-0.37957	0.394327	0.401056	0.406231	58.8637902	394.5702

Figure 13-18. *Only layer 1 parameters are optimized*

The transformed data of layer 1 is the input for layer 2 neurons. Follow these instructions to complete layer 2 input aggregation and data transformation:

14. Merge cells R9:T9; enter "2nd hidden layer aggregation" inside the merged cell.

15. Enter numbers 1, 2, and 3 in cells R10, S10, and T10, respectively.

16. In cell R11, enter the following formula:

```
=SUMPRODUCT($M11:$O11,INDEX($I$2:$K$4,R$10, 0))+
INDEX($L$2:$L$4,R$10,1)
```

This formula aggregates layer 1's transformed data by using the function SUMPRODUCT. Its mathematical equation is similar to $y = w1x2 + w2x2 + w3x3 + b$.

17. Autofill cell R11 to cell T11, and then autofill together to cells R182:T182.

18. We are going to make use of sigmoid function again to transform the layer 2 data. However, we do not want to normalize the data again. We can certainly normalize the data again into range [-1, 1] before applying sigmoid transformation if we so desired. Merge cells U9:W9; enter "2nd layer transformation" in the merged cell.

19. Enter numbers 1, 2, and 3 in cells U10, V10, and W10, respectively.

20. Enter the formula =1/(1+EXP(-R11)) in cell U11, autofill to cell W11, and then autofill together to cells U182:W182.

21. Enter "Output-2" in cell X10.

22. Enter "Error-2" in cell Y10.

Part of our worksheet looks like Figure 13-19.

▲	M	N	O	P	Q	R	S	T	U	V	W	X	Y
1	Wo2	Bo2		Error Sum1	26765								
2	1	1											
3	1.5	1.5											
4	2	2											
5													
6													
7													
8													
9	sigmoid transformed					2nd hidden layer aggregation			2nd layer transformation				
10	1	2	3	Output-1	Error-1	1	2	3	1	2	3	Output-2	Error-2
11	0.36845	0.369751	0.358591	55.1051516	259.3759	2.096792	3.145188	4.193583	0.890591	0.958719	0.985132		
12	0.375239	0.384342	0.378164	56.6488278	311.4811	2.137745	3.206617	4.275489	0.894518	0.961083	0.986285		
13	0.432925	0.406279	0.397585	59.088838	403.5614	2.23679	3.355184	4.473579	0.903505	0.966274	0.988722		
14	0.479079	0.437448	0.422878	61.9528035	526.8312	2.339405	3.509107	4.678809	0.912088	0.970946	0.990795		
15	0.355034	0.341249	0.328783	52.5265983	182.9689	2.025066	3.037599	4.050132	0.883404	0.954244	0.982878		
16	0.394327	0.401056	0.406231	58.8637902	394.5702	2.201614	3.302422	4.403229	0.900394	0.964512	0.98791		
17	0.461211	0.46218	0.465414	64.6606655	658.4698	2.388804	3.583206	4.777608	0.91597	0.972965	0.991654		
18	0.387402	0.374014	0.359611	55.5402921	273.5813	2.121027	3.181541	4.242054	0.89293	0.960134	0.985826		
19	0.462905	0.439026	0.433988	62.3601999	545.6989	2.33592	3.503879	4.671839	0.911809	0.970798	0.990732		

Figure 13-19. *Layer 2 data transformation without normalization*

The following instructions guide us how to continue:

23. In cell X11, enter the formula =MMULT(U11:W11,$M$2:$M$4)+$N$2 to generate output.

24. Autofill from X11 to X182.

25. In cell Y11, enter the formula =(F11-X11)^2 to calculate the error.

26. Autofill from Y11 to Y182.

27. In cell S1, enter "Error Sum2".

28. In cell T1, enter the formula =SUM(Y11:Y182).

Part of our worksheet looks like Figure 13-20.

▲	N	O	P	Q	R	S	T	U	V	W	X	Y
1	Bo2		Error Sum1	26765		Error Sum2	737862.9					
2	1											
3	1.5											
4	2											
5												
6												
7												
8												
9	оid transformed				2nd hidden layer aggregation			2nd layer transformation				
10	2	3	Output-1	Error-1	1	2	3	1	2	3	Output-2	Error-2
11	0.369751	0.358591	55.1051516	259.3759	2.096792	3.145187577	4.193583	0.890591	0.958719	0.985132	5.298934	1135.762
12	0.384342	0.378164	56.6488278	311.4811	2.137745	3.206616929	4.275489	0.894518	0.961083	0.986285	5.308713	1135.103
13	0.406279	0.397585	59.088838	403.5614	2.23679	3.355184296	4.473579	0.903505	0.966274	0.988722	5.330361	1133.645
14	0.437448	0.422878	61.9528035	526.8312	2.339405	3.509106953	4.678809	0.912088	0.970946	0.990795	5.350098	1132.316
15	0.341249	0.328783	52.5265983	182.9689	2.025066	3.037599341	4.050132	0.883404	0.954244	0.982878	5.280526	1137.003
16	0.401056	0.406231	58.8637902	394.5702	2.201614	3.302421672	4.403229	0.900394	0.964512	0.98791	5.322982	1134.142
17	0.46218	0.465414	64.6606655	658.4698	2.388804	3.583206191	4.777608	0.91597	0.972965	0.991654	5.358725	1131.735
18	0.374014	0.359611	55.5402921	273.5813	2.121027	3.181540602	4.242054	0.89293	0.960134	0.985826	5.304782	1135.368

Figure 13-20. *Errors calculated*

29. Let's use Solver again to optimize all the weights and biases. Click Data ➤ Solver to start the Solver and set up the Solver parameters as shown in Figure 13-21.

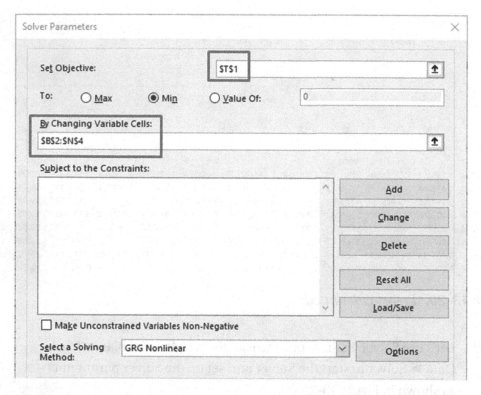

Figure 13-21. *Use Solver again to optimize the parameter set of all layers*

After the optimization of the weights and biases by Solver, part of our worksheet looks like Figure 13-22. Since the model becomes more complicated, your result may look slightly different from that in Figure 13-22. Don't be surprised.

	W21	W22	W23	B2	Wo2	Bo2		Error Sum1	28342.09		Error Sum2	21256.68
1	W21	W22	W23	B2	Wo2	Bo2		Error Sum1	28342.09		Error Sum2	21256.68
2	1.971081	1.98513	2.088216	4.053192	14.69958	15.82929						
3	0.335534	0.318302	0.336957	-3.52748	32.41488	1.5						
4	4.795747	4.99863	4.75973	-6.97899	56.14744	2						
5												
6	441.7852											
7	167.7952											
8												
9	d sum	normalized weighted sum			sigmoid transformed					2nd hidden layer aggregation		
10	3	1	2	3	1	2	3	Output-1	Error-1	1	2	3
11	244.4329	-0.00126	-0.36673	-0.44058	0.499686	0.409331	0.391603	59.7681188	431.3148	6.66844	-3.09757101	-0.6726
12	254.6842	0.070005	-0.26305	-0.36575	0.517494	0.434613	0.409568	61.6935929	514.9992	6.791245	-3.07749488	-0.37531
13	285.6565	-0.43722	-0.26645	-0.13967	0.392404	0.433778	0.46514	62.8764668	570.0857	6.65907	-3.10100733	-0.71488
14	319.7066	-0.30616	-0.08091	0.108883	0.424053	0.479784	0.527194	67.8995512	835.1841	6.942363	-3.05483473	-0.03777
15	236.6426	-0.35092	-0.52192	-0.49745	0.41316	0.372403	0.378141	56.8051818	317.0245	6.396471	-3.14289383	-1.33623
16	247.8712	-0.46696	-0.39571	-0.41548	0.385337	0.402343	0.397598	58.3310039	373.6877	6.441696	-3.13614312	-1.22739

Figure 13-22. Optimize all parameters by Solver

As usual, we now need to set up the cutoff. Follow these instructions:

30. Enter "mean" in cell S6 and "count" in cell S7.

31. Enter numbers 39, 69, and 84 in cells T5, U5, and V5.

32. Enter "cutoff" in cell W5 and "missed" in cell X5.

33. Enter the formula =AVERAGEIFS($X$11:$X$182,$F$11:$F$182,T$5) in cell T6. This formula calculates the Output mean for those samples whose Class values equal to 39. Autofill from T6 to V6.

34. Enter the formula =COUNTIFS($F$11:$F$182,T$5) in cell T7. This counts the number of samples whose Class values equal to 39. Autofill from T7 to V7.

35. Enter the formula =(T6*T7+U6*U7)/(T7+U7) in cell W6. This is the cutoff between the two classes 39 and 69.

36. Enter the formula =(U6*U7+V6*V7)/(U7+V7) in cell W7. This is the cutoff to differentiate the two classes 69 and 84.

Part of our worksheet looks exactly like Figure 13-23.

◢	R	S	T	U	V	W	X	Y
1		Error Sum2	21256.68					
2								
3								
4								
5			39	69	84	cutoff	missed	
6		mean	51.06138	65.27256	79.63806	58.98972		
7		count	42	53	77	73.78136		
8								
9	2nd hidden layer aggregation			2nd layer transformation				
10	1	2	3	1	2	3	Output-2	Error-2
11	6.66844	-3.09757101	-0.6726	0.998731	0.043208	0.337914	50.8838	141.2248
12	6.791245	-3.07749488	-0.37531	0.998878	0.044045	0.407258	54.80656	249.8475
13	6.65907	-3.10100733	-0.71488	0.998719	0.043066	0.328521	50.35163	128.8595
14	6.942363	-3.05483473	-0.03777	0.999035	0.045009	0.490557	59.5172	420.9553
15	6.396471	-3.14289383	-1.33623	0.998335	0.041372	0.208131	43.53152	20.53471

Figure 13-23. *Cutoff for the neural network model with two hidden layers*

We want to use the training data to estimate how well our model is. As our data size is not large, we are not performing a cross-validation here.

37. Enter "Predict" in cell Z10 and "Diff" in cell AA10.

38. Enter the formula =IF(X11<$W$6,39,IF(X11<$W$7,69,84)) in cell Z11. This formula categorizes each sample based on the Output value and the two cutoff values. Autofill from Z11 to Z182.

39. Enter the formula =IF(F11=Z11,0,1) in cell AA11. This formula returns 0 if the predicted class type is the same as the existing class type; otherwise, it returns 1. Autofill from AA11 to AA182.

40. Enter the formula =SUM(AA11:AA182) in cell X6. This formula counts how many predicted class types are incorrect.

Our result should be similar as Figure 13-24. Our neural network with two hidden layers incorrectly predicted 48 out of 172 samples.

▲	R	S	T	U	V	W	X	Y	Z	AA
1		Error Sum2	21256.68							
2										
3										
4										
5			39	69	84	cutoff	missed			
6		mean	51.06138	65.27256	79.63806	58.98972	48			
7		count	42	53	77	73.78136				
8										
9	2nd hidden layer aggregation			2nd layer transformation						
10	1	2	3	1	2	3	Output-2	Error-2	Predict	Diff
11	6.66844	-3.09757101	-0.6726	0.998731	0.043208	0.337914	50.8838	141.2248	39	0
12	6.791245	-3.07749488	-0.37531	0.998878	0.044045	0.407258	54.80656	249.8475	39	0
13	6.65907	-3.10100733	-0.71488	0.998719	0.043066	0.328521	50.35163	128.8595	39	0
14	6.942363	-3.05483473	-0.03777	0.999035	0.045009	0.490557	59.5172	420.9553	69	1
15	6.396471	-3.14289383	-1.33623	0.998335	0.041372	0.208131	43.53152	20.53471	39	0

Figure 13-24. *The neural network incorrectly predicted 48 samples*

You may obtain a result slightly different from what is shown in Figure 13-24. If so, don't be surprised. This difference is caused by Solver. We have been experiencing Solver in learning data mining through Excel. Solver is a mathematical optimization tool that can sometimes generate slightly different results due to its internal algorithm. If your Excel is using a newer version of Solver, your model may have a better performance.

The next step in this experiment is to make prediction on the scoring dataset. Follow these instructions:

41.   Select cells G182:X182. Autofill to cells G190:X190.

42.   Autofill from cell Z182 to Z190.

The predicted results look similar to Figure 13-25.

⃟	A	B	C	D	E	X	Y	Z	AA
175		50.9	-7.4	-2.4	7	82.1449	3.441381	84	0
176		50.9	-5.4	-0.6	7	84.15497	0.024017	84	0
177		46.1	-6.7	-2.6	7	72.99528	121.1039	69	1
178		46.1	-5	-0.6	8	80.0014	15.98878	84	0
179		50.9	-4.9	-2.7	8	74.86475	83.4528	84	0
180		46.1	-4.8	-0.6	7	78.79261	27.11689	84	0
181		60.4	-5.4	-2.4	7	85.27806	1.633437	84	0
182		55.6	-5.1	-2.7	7	80.73599	10.65379	84	0
183	Scoring data set	47.6	-5.9	-4.3	6	56.93885		39	
184		38.1	-3.8	-1.8	7	49.68801		39	
185		47.6	-5.9	-4.6	5	53.08293		39	
186		52.4	-5.6	-3.7	5	69.80611		69	
187		47.6	-6.4	-2.5	7	75.10016		84	
188		52.4	-5.8	-3.6	8	74.6359		84	
189		57.6	-6.4	-1.5	8	86.52697		84	
190		42.9	-3.3	-2.4	6	51.12411		39	
191									

Figure 13-25. *The predicted class types for the scoring dataset*

We may have quite a few questions about the preceding experiment process. For example, we may ask why we need to generate output-1, error-1, output-2, and error-2 to optimize layer 1 and layer 2 parameters separately. That is a good question. We can certainly practice this neural network model a little differently, for example, without optimizing layer 1 parameters first. In fact, we might even get a better result. This is the advantage of using Excel to learn data mining methods, especially those complex ones. Because we have a clear image about every step, we may raise various questions in the learning process and some of them may help reform and improve certain data mining algorithms.

We can also make use of a different transformation function in hidden layer 2. It would be easy to test in Excel if using a different transformation function (such as the rectified linear unit function) in a different layer can improve the model result. This is a good exercise to try out.

This wraps up the study of artificial neural networks in Excel. Please find the final result in Chapter13-2b.xlsx.

Reinforcement Exercises

Though there are three reinforcement exercises, they are all based on the data we used in Experiment 2.

1. Exercise 1: Start with Chapter13-HW-1.xlsx which is the same as Chapter13-2a.xlsx; please construct a 2-hidden-layers neural network as shown in Experiment 2. However, when using Solver to minimizing Error Sum2, please optimize parameters in $H$2:$N$4 only. Compare the prediction result with that of Experiment 2. Refer to Chapter13-HW-1-withAnswers.xlsx.

2. Exercise 2: Start with Chapter13-HW-2.xlsx which is also the same as Chapter13-2a.xlsx; please construct a 2-hidden-layers neural network as shown in Experiment 2. However, do not optimize layer 1 parameters first. Instead, optimize all parameters together after constructing both hidden layers. Compare the prediction result with that of Experiment 2. Refer to Chapter13-HW-2-withAnswers.xlsx.

3. Exercise 3: Start with Chapter13-HW-3.xlsx which makes use of four neurons. Please construct a neural network with one or two hidden layers. Compare the prediction result with that of Experiment 2. Refer to Chapter13-HW-3-withAnswers.xlsx.

It would be interesting to compare the results obtained from Experiment 2, Exercise 1, Exercise 2, and Exercise 3.

Review Points

1. General understanding of neural network

2. Input layer, hidden layer(s), and output layer

3. Aggregation function, transformation function, or activation function, especially sigmoid function

4. Excel functions SUMPRODUCT, COUNTIFS, INDEX, AVERAGEIFS, MMULT, and EXP

5. What is deep learning

6. Solver

Text Mining

Text mining is such a large topic in data mining that it can be a separate field. It is also so important and has so many applications that a typical data mining book will never miss. So far, all the data we have been using are structured data. By definition, structured data indicate that the data are arranged in a specific format such that they can be easily processed by computer programs. The data we have been using have already been arranged in a table format, a format that is requested by all the data mining methods we have learned. However, text data, what text mining methods work with, is a form of unstructured data and is usually sequential data. Other forms of unstructured data include audio files, images, etc.

Is it possible to apply techniques we have learned to mine unstructured forms of data like articles or web pages? The answer is yes. There is a prerequisite, however.

General Understanding

To apply the data mining techniques we have learned to unstructured text data, the data must be preprocessed to be structured or at least semistructured. There are several typical techniques to preprocess (prepare) text data. Such techniques include case transformation, tokenization, stop-word removal, stemming, generating n-grams, token replacement, etc. After the text data are organized in a table format, we can build up a term-document matrix. From there, meaningful information can be mined including term frequency and term pattern for various applications, such as risk management, customer care service, fraud prevention, spam filtering, social media text analysis, etc.

Assume that we are reading a large number of reviews about a hotel. Our purpose is to grade the hotel based on the number of positive comments vs. the number of negative comments. The first task is to determine what is a positive or negative review. Such a task is similar to sentence sentiment analysis: determining the emotion or sentiment of a short sentence. The simplest technique to assess the sentiment of a sentence is

© Hong Zhou 2023
H. Zhou, *Learn Data Mining Through Excel*, https://doi.org/10.1007/978-1-4842-9771-1_14

counting the appearances of positive keywords and negative keywords. Certainly, there should be a dictionary of such keywords (and phrases), representing either positive or negative emotions. MPQA Subjectivity Lexicon is exactly such a dictionary and it is adopted by Excel Azure machine learning add-in. In MPQA Subjectivity Lexicon, there are about 8000 words; each is assigned a strong or weak polarity score. Since this book does not use add-ins except for the built-in add-in Solver, we are not going to use Azure or MPQA. Instead, I am going to illustrate how to conduct a simple opinion (sentiment) analysis by creating a small dictionary in Excel. Let's go through the data preparation steps first by working on one review "The hotel it is fantastic".

1. Case transformation: It is a good idea to transform all words into lowercase. We do not want to treat "Fantastic" differently from "fantastic". After case transformation, the original review becomes "the hotel it is fantastic".

2. Tokenization: By tokenization, we separate each word into an independent term. In the process of tokenization, we also remove punctuations and white spaces. After tokenization, the original sentence becomes five separate words: the, hotel, it, is, fantastic.

3. Stop words are those that do not have actual meaning in the context. They are necessary conjunctions and articles in English, for example, then, a, the, it, is, of, for, just, to, etc. Removing them can simplify the texts. After removing stop words, we have only two words left: hotel and fantastic.

4. Assume that "fantastic" is a positive word in our dictionary with polarity score 1 and "hotel" is a neutral word with polarity score 0. Thus, this review is assessed as positive (score = 1).

If the review becomes "The hotel it is not fantastic" instead, we know that it should not be a positive comment even though the keyword fantastic is there. How to assess this review to be negative? There can be multiple ways.

- Way 1: Mark "not" as a negative keyword. If we take this approach, what about a review like "the hotel is not bad"?

- Way 2: Take "not fantastic" or "not bad" as a phrase and assess their meaning based on the phrase. This is what n-gram technique means.

- Way 3: Combine way 1 and way 2 together. Mark "not" as a negative keyword and take "not fantastic" or "not bad" as a phrase.

In text mining, an n-gram is a combination of n words. A single word can miss some meaning. Good is the opposite of "not good" and different in intensity from "very good". Applying grams can make our text mining activities more granular but can also incur much more complexity. In this chapter, we will only consider 2-grams for the purpose of simplicity. This indicates that we will consider "not fantastic" or "not bad" as a phrase.

We didn't make use of the stemming technique in the preceding example. Instead of writing "The hotel it is fantastic", people may write it as "The hotel it is a fantasy". Both "fantasy" and "fantastic" have the same word origin. To reduce the text mining complexity, the stemming technique would treat "fantasy" as "fantastic". Another example would be to treat "thanks" the same as "thank".

Learn Text Mining Through Excel

Let's switch to Excel to exercise one text mining example. Open the file Chapter14-1a. xlsx. There are two worksheets there. The worksheet named "dictionary" contains some stop words and some positive vs. negative keywords (called Polarity Words in the worksheet). Note, these words are specifically collected for this text mining demonstration only. Figure 14-1 shows part of the worksheet "dictionary".

	A	B	C	D
1	**Polarity Words**	**Polarity Score**		**Stop Words**
2	not	-1		a
3	amazing	1		an
4	best	1		and
5	cozy	1		are
6	delightful	1		as
7	enjoy	1		at
8	excellent	1		can
9	fantastic	1		could
10	good	1		for
11	lovely	1		in
12	memorable	1		is
13	outstanding	1		it
14	pleasant	1		of
15	pleasure	1		or
16	wonderful	1		that
17	thank	1		the
18	come back	1		then
19	happy	1		this
20	not amazing	-1		thus
21	not best	-1		to
22	not cozy	-1		was

Figure 14-1. *Some polarity keywords and their negations, and some stop words*

As shown in Figure 14-1, column D in worksheet dictionary stores some stop words. Both positive and negative single keywords together with their negated 2-gram phrases are stored in column A. Column B keeps the scores associated with each keyword. Note "not" is designated as a negative keyword (cell A2) with score -1. Except for the keyword "not", all other single keywords in Figure 14-1 are positive. Their negated 2-gram phrases are all assigned a score of -1. Why? Let me use the short sentence "It is not amazing" as an example to explain. Given the short sentence "It is not amazing", the two single keywords "not" and "amazing" would generate -1 and 1, respectively. The 2-gram phrase "not amazing" gives -1; therefore, the total score of this short sentence equals $-1 + 1 - 1 = -1$: negative.

	A	B
37	bad	-1
38	difficult	-1
39	expensive	-1
40	outrageous	-1
41	ridiculous	-1
42	rude	-1
43	sick	-1
44	terrible	-1
45	unhappy	-1
46	unpleasant	-1
47	not bad	2
48	not difficult	2
49	not expensive	2
50	not outrageous	2
51	not ridiculous	2
52	not rude	2
53	not sick	2
54	not terrible	2
55	not unhappy	2
56	not unpleasant	2

Figure 14-2. *Single negative keywords and their negated 2-gram phrases*

The scores assigned to the 2-gram phrases of single negative keywords are all 2, as shown in Figure 14-2. For example, "not bad" has a score of 2. Why? Consider the short sentence "It is not bad". Both "not" and "bad" return -1, while "not bad" returns 2. Therefore, the total score of this short sentence is $-1 - 1 + 2 = 0$: neutral.

Remember that the preceding score assignments are for this text mining task only. We can certainly modify the "dictionary" based on other preferences. For example, if we prefer "not bad" to be positive, then we may want to assign the 2-gram phrase "not bad" a score of 3. The scoring system used here is just an example.

To make easy use of these keywords and stop words, we would like to name them. Click the main tab Formula ➤ click Name Manager on the upcoming menu ➤ click the button New. This procedure is explained in Figure 14-3. Note that Name Manager is available since Excel 2010.

257

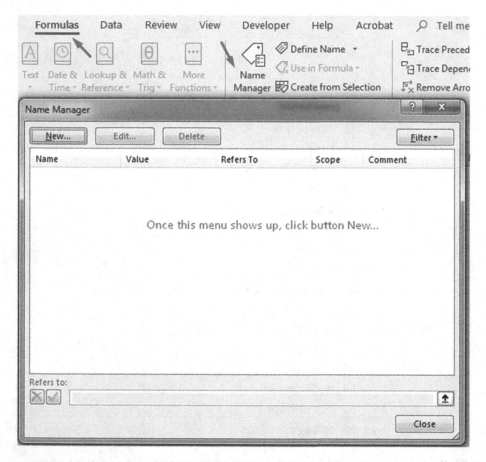

Figure 14-3. *Create names for keywords and stop words*

Clicking the button "New..." as shown in Figure 14-3 would pop up another small menu, as illustrated in Figure 14-4. As the keywords are in the array A2:A56, the "Refers to:" is set to `=dictionary!$A$2:$A$56`. Enter "kw" for the name.

Be aware that the "Scope" is Workbook. This indicates that the name kw is accessible in every worksheet. Click OK.

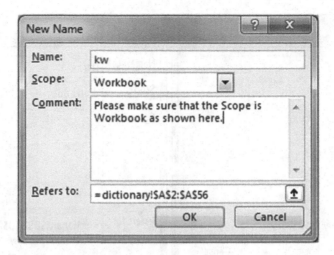

Figure 14-4. *Name the array for keywords*

After clicking the button OK, the menu in Figure 14-3 is updated to be like Figure 14-5.

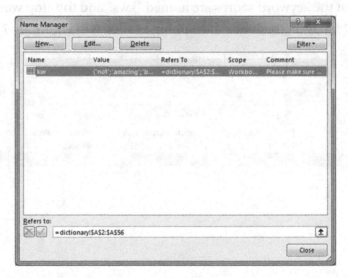

Figure 14-5. *After keyword array is named kw*

Continue to name the array of keyword scores (called "Polarity Scores" in the worksheet) and the array of stop words, respectively, as shown in Figure 14-6.

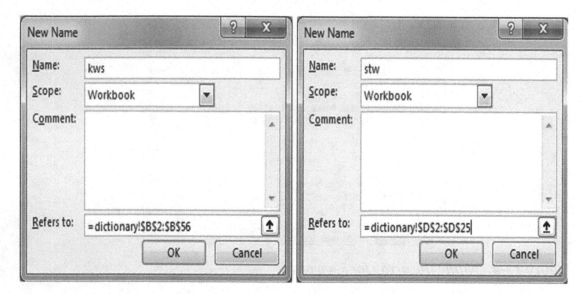

Figure 14-6. *Name the keyword scores and stop words*

Make sure that the keyword scores are named "kws" and the stop words are named "stw". The Name Manager concludes the naming as shown in Figure 14-7.

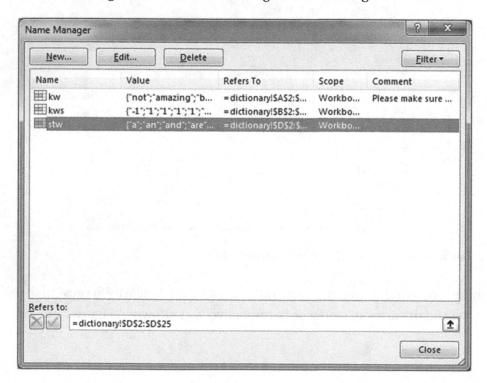

Figure 14-7. *Confirm the naming*

By naming those arrays, we can now reference the keywords by the name "kw", keyword scores by "kws", and stop words by "stw".

There are 34 reviews in the worksheet named "hotelReview". They are placed in cells A1:A34. Take a close look at them. Figure 14-8 displays part of the reviews.

	A
1	The hotel it is fantastic. Environment it is great !
2	Cozy
3	Breakfast has high-quality products, but it lacks in choice.
4	Price utterly ridiculous for what you get.
5	What a wonderful hotel
6	not bad at all.
7	We will come back!
8	One of the best Hotels.
9	10 stars!
10	A peaceful and amazing place.
11	The shower did not have hot water! Called to get fixed.
12	Everything OK.
13	We had a wonderful experience. Very sad we have to leave.

Figure 14-8. *A glance of the reviews*

Follow these steps to complete our text mining exercise:

1. As we have learned, we need to transform all the words into lowercase and trim the spaces around each sentence. Enter the formula =TRIM(LOWER(A1)) in cell B1, and then autofill to cell B34. This formula transforms the text into lowercase and trims off both the leading and tailing blank spaces.

2. Some sentences are ended by exclamation. To remove exclamations, enter the following formula in cell C1:

   ```
   =IF(MID(B1,LEN(B1),1)="!", TRIM(LEFT(B1,LEN(B1)-1)),B1)
   ```

   LEN(B1) returns the length of the text inside cell B1.
   MID(B1,LEN(B1),1) picks the last character of the text in B1. If the last character is "!", LEFT(B1,LEN(B1)-1) removes "!" by taking all the characters except the last one. If the last character is not "!", it takes whatever is inside B1.

   Note: after removing "!", TRIM(LEFT(B1,LEN(B1)-1)) trims any possible spaces again.

3. Before we tokenize the words, we need to copy the sentences in column C into column D via Paste Special (paste values). If we don't, the Excel feature "Text to Columns" we are going to employ will split the formulas instead of the text values in column C. Let's copy and paste cells C1:C34 to column D by values. Part of our worksheet looks like Figure 14-9.

	A	B	C
1	The hotel it is fantastic. Environment it is great !	the hotel it is fantastic. environment it is great !	the hotel it is fantastic. environment it is great
2	Cozy	cozy	cozy
3	Breakfast has high-quality products, but it lacks in choice.	breakfast has high-quality products, but it lacks in choice.	breakfast has high-quality products, but it lacks in choice.
4	Price utterly ridiculous for what you get.	price utterly ridiculous for what you get.	price utterly ridiculous for what you get.
5	What a wonderful hotel	what a wonderful hotel	what a wonderful hotel
6	not bad at all.	not bad at all.	not bad at all.
7	We will come back!	we will come back!	we will come back
8	One of the best Hotels.	one of the best hotels.	one of the best hotels.

Figure 14-9. *Worksheet after paste values*

4. The tool tab Text to Columns is under the main tab Data in Excel. Once clicking the main tab Data, the tool tab Text to Columns shows up. Refer to Figure 14-10.

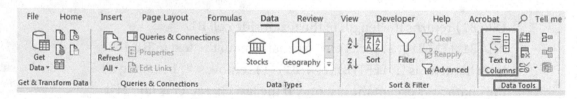

Figure 14-10. *The tool tab Text to Columns is under the main tab Data*

5. Select column D and click the tool tab Text to Columns; a menu shows up. Choose "Delimited" as shown in Figure 14-11.

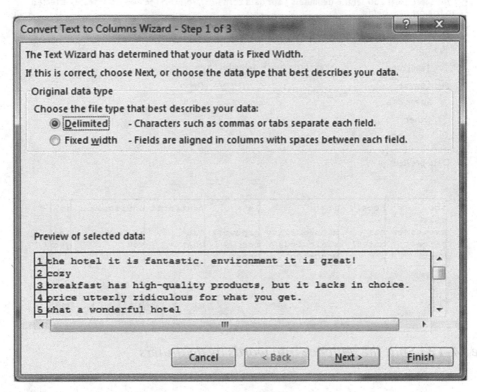

Figure 14-11. *Choose delimited and click Next*

6. Click Next; on the next menu, select all delimiters and enter period "." in the box next to "Other", as illustrated in Figure 14-12.

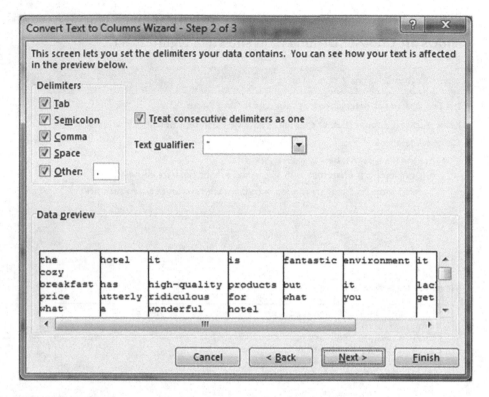

Figure 14-12. *Split the sentence into words by delimiters*

7. On the next menu, click Finish as shown in Figure 14-13. Note:
    the destination is cell D1. By selecting D1 as the destination, the
    tokenized words will be placed in an area starting from column
    D. This indicates that the original texts inside column D will be
    replaced.

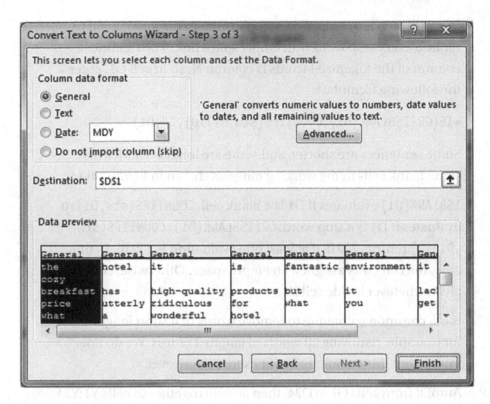

Figure 14-13. *Destination is D1*

By now, part of our worksheet looks like Figure 14-14.

⊿	D	E	F	G	H	I	J	K	L
1	the	hotel	it	is	fantastic	environm	it	is	great
2	cozy								
3	breakfast	has	high-quali	products	but	it	lacks	in	choice
4	price	utterly	ridiculous	for	what	you	get		
5	what	a	wonderfu	hotel					
6	not	bad	at	all					
7	we	will	come	back					
8	one	of	the	best	hotels				

Figure 14-14. *A glance of the tokenized words*

8. Our next step is to remove the stop words. To do so, we need to replace every stop word with empty space first. The rightmost column of the tokenized words is column N, so in cell O1, enter the following formula:

```
=IF(OR(ISBLANK(D1),COUNTIFS(stw,D1)>0),"",D1)
```

Some sentences are shorter, and some are longer. This leaves some blank cells in the working range as shown in Figure 14-14.

ISBLANK(D1) evaluates if D1 is a blank cell. COUNTIFS(stw,D1)>0 evaluates if D1 is a stop word. OR(ISBLANK(D1),COUNTIFS(stw,D1)>0) returns TRUE if at least one evaluation is TRUE. In this case, the cell O1 is assigned an empty space. Otherwise, cell O1 is given whatever inside cell D1.

It is a common technique to remove words of a short length, for example, removing all words of length 3 or less. We do not implement this technique in our exercise, however.

9. Autofill from cells O1 to O34, then autofill together to cells Y1:Y34. Part of our worksheet looks like Figure 14-15.

	O	P	Q	R	S	T	U	V	W
1		hotel			fantastic	environm			great
2	cozy								
3	breakfast	has	high-quali	products	but		lacks		choice
4	price	utterly	ridiculous		what	you	get		
5	what		wonderfu	hotel					
6	not	bad		all					
7			come	back					
8	one			best	hotels				

Figure 14-15. *Stop words are replaced by empty space*

10. To remove the stop words, our plan is to make blank the cells containing stop words, then concatenate remaining words for each sentence, and then apply Text to Columns to retokenize. Leave column Z untouched. In cell AA1, enter the following formula:

```
=IF(ISBLANK(O1)=FALSE,TRIM(Z1 & " " & O1), TRIM(Z1 & ""))
```

When we autofill this formula horizontally, it accumulates the texts in cells but skips blank cells. As column Z is left untouched, cells in column Z are all blank cells. If `ISBLANK(O1)=FALSE`, that is, cell O1 is not a blank cell, concatenate the cell before AA1 (Z1) with O1. Otherwise, only keep in cell AA1 whatever the cell before AA1 has.

Since Office 2019 or Office 365, Excel has a new function named TEXTJOIN which can

- Concatenate the texts from multiple ranges

- Ignore empty cells

- And insert a desired delimiter

The function TEXTJOIN is definitely a much better choice than the preceding formula if it is available in your Excel. Assume that it is available, then

- Ignore the preceding formula.

- Enter the formula `=TEXTJOIN(" ",TRUE,O1:Y1)` in cell AK1.

- Autofill from cell AK1 to cell AK34.

- And skip the following step (step 11).

11. Autofill from cell AA1 to cell AK1 and then autofill together to cells AA34:AK34. Column AK keeps the sentences without stop words. Check Figure 14-16.

◢	AK
1	hotel fantastic environment great
2	cozy
3	breakfast has high-quality products but lacks choice
4	price utterly ridiculous what you get
5	what wonderful hotel
6	not bad all
7	come back
8	one best hotels

Figure 14-16. *After removing stop words*

12. Copy cells AK1:AK34 and paste values only to column AL.

13. Use tool tab Text to Columns to tokenize the texts inside column AL. Refer to steps 4–7. Part of our worksheet looks like Figure 14-17.

◢	AL	AM	AN	AO	AP	AQ	AR	AS	AT
1	hotel	fantastic	environment	great					
2	cozy								
3	breakfast	has	high-quality	products	but	lacks	choice		
4	price	utterly	ridiculous	what	you	get			
5	what	wonderful	hotel						
6	not	bad	all						
7	come	back							
8	one	best	hotels						
9		10	stars						
10	peaceful	amazing	place						
11	shower	did	not	have	hot	water!	called	get	fixed

Figure 14-17. *Retokenized words*

14. To generate 2-gram phrases, enter the following formula in cell AU1:

```
=IF(AND(ISBLANK(AL1)=FALSE, ISBLANK(AM1)=FALSE),AL1 & " "
& AM1,"")
```

This formula guarantees that only when both cells AL1 and AM1 are not blank cells, a 2-gram phrase is created inside cell AU1.

15. Autofill from cell AU1 to BB1 and then autofill together from cells AU1:BB1 to cells AU34:BB34.

   Our 2-gram phrases look like Figure 14-18. Note there is no 2-gram phrase for row 2. Also note the last 2-gram phrase is in column BB (not shown in Figure 14-18).

◢	AU	AV	AW	AX
1	hotel fantastic	fantastic environment	environment great	
2				
3	breakfast has	has high-quality	high-quality products	products but
4	price utterly	utterly ridiculous	ridiculous what	what you
5	what wonderful	wonderful hotel		
6	not bad	bad all		
7	come back			
8	one best	best hotels		

Figure 14-18. *2-gram phrases*

16. All words are tokenized in cells AL1:BB34, including both single words and 2-gram phrases. It is time to calculate the scores for each sentence. In cell BC1, enter the following formula:

```
=IFERROR(INDEX(kws, MATCH(AL1, kw, 0), 1), 0)
```

This formula calculates the polarity score for the word in cell AL1 ("hotel").

When the MATCH function cannot find an exact match of AL1 in the keyword array (represented by the name kw), it gives a #NA error. Thus, the IFERROR function guarantees when there is no such keyword and therefore a #NA error occurs, 0 is returned. When MATCH function succeeds, it returns the word position in the array kw. This position is fed to the function INDEX to find the corresponding score in the keyword score array (represented by kws).

Pay close attention to the expression INDEX(kws, MATCH(AL1, kw, 0), 1). The last parameter inside MATCH function is 0, indicating an exact match is demanded. The last parameter inside the INDEX function is 1 since kws has only one column.

17. Autofill from cell BC1 to cell BS1, then autofill together to cells BC34:BS34. BC1:BS34 is a sparse matrix. Part of our worksheet looks like Figure 14-19.

	A	BC	BD	BE	BF	BG	BH
1	The hotel it is fantastic. Environment it is great !	0	1	0	0	0	0
2	Cozy	1	0	0	0	0	0
3	Breakfast has high-quality products, but it lacks in choice.	0	0	0	0	0	0
4	Price utterly ridiculous for what you get.	0	0	-1	0	0	0
5	What a wonderful hotel	0	1	0	0	0	0
6	not bad at all.	-1	-1	0	0	0	0
7	We will come back!	0	0	0	0	0	0
8	One of the best Hotels.	0	1	0	0	0	0
9	10 stars!	0	0	0	0	0	0
10	A peaceful and amazing place.	0	1	0	0	0	0
11	The shower did not have hot water! Called to get fixed.	0	0	-1	0	0	0
12	Everything OK.	0	0	0	0	0	0
13	We had a wonderful experience. Very sad we have to leave.	0	1	0	0	0	0
14	Thanks for the stay in this wonderful Hotel.	0	0	1	0	0	0
15	A five star stay!	0	0	0	0	0	0

Figure 14-19. *Polarity scores computed for tokenized words and 2-grams*

18. In cell BT1, enter the formula =SUM(BC1:BS1). This formula sums up all the scores for the first sentence.

19. Autofill from cell BT1 to cell BT34.

20. In cell BU1, enter the formula:

    =IF(BT1>0,"Positive", IF(BT1=0,"Neutral","Negative"))

    If the summed score is positive, the sentence is considered positive. If the summed score is 0, the sentence is considered neutral, otherwise negative.

21. Autofill from cell BU1 to cell BU34.

22. By freezing column B (click column B ➤ click View ➤ click Freeze Panes ➤ select Freeze Panes), our final result can be viewed as Figure 14-20.

◢	A	BT	BU
1	The hotel it is fantastic. Environment it is great !	1	Positive
2	Cozy	1	Positive
3	Breakfast has high-quality products, but it lacks in choice.	0	Neutral
4	Price utterly ridiculous for what you get.	-1	Negative
5	What a wonderful hotel	1	Positive
6	not bad at all.	0	Neutral
7	We will come back!	1	Positive
8	One of the best Hotels.	1	Positive
9	10 stars!	0	Neutral
10	A peaceful and amazing place.	1	Positive
11	The shower did not have hot water! Called to get fixed.	-1	Negative
12	Everything OK.	0	Neutral
13	We had a wonderful experience. Very sad we have to leave.	1	Positive
14	Thanks for the stay in this wonderful Hotel.	1	Positive
15	A five star stay!	0	Neutral
16	Come back again is not even a question.	0	Neutral
17	We had a truly memorable and delightful stay. Thank you all.	3	Positive
18	People here are really nice. Enjoy.	1	Positive
19	A lovely and warm place	1	Positive

Figure 14-20. Sentiment analysis of hotel reviews

The complete text mining work is available in Chapter14-1b.xlsx and Chapter14-1c.xlsx.

Chapter14-1b.xlsx does not make use of the function TEXTJOIN, while Chapter14-1c.xlsx does.

A careful examination of our result reveals that our text mining model does not work well for reviews 9, 15, 16, 20, and 26. This does not mean our mining job is a failure. Reviews 9 ("10 stars!"), 15 ("A five star stay!"), and 20 ("Not cheap, but worthy.") can be properly assessed by adding more keywords. Review 26 ("want to say 'Thanks' to all of you!") can also be properly assessed by applying the stemming technique.

Some words imply the same meaning though they are different in writing, for example, "thanks" and "thank you". Other examples could be "book" and "books", "imply" and "implies", etc. "Thanks", "books", and "implies" are all derived from their basic stems ("thank", "book", and "imply"). Reducing words to their basic stems is called stemming in text mining. Certainly, reducing "thanks" to "thank" can assess review 26 to be positive.

What is really challenging is review 16 ("Come back again is not even a question."). We understand this is a positive comment, but many computer algorithms will fail to recognize it as such. It is because of sentences like this that text mining is a very challenging and complex task. This chapter only provides a learning and education example. Therefore, I won't explain how to handle such a sentence in this book.

Text mining can also identify the words that appear the most frequently in reviews, comments, and customer support inquires. Words of high term frequency can inform us of the common issues that customers are facing. Term frequency (TF) can also be used together with another concept, inverse document frequency (IDF), to rank the web pages in a web search.

Certainly, Excel is not suitable for ranking web pages. This is one example showing the limitation of applying Excel in data mining. However, Excel is becoming more and more powerful in data mining or machine learning by more and more add-ins, though this book tries to demonstrate the use of Excel without additional add-ins.

Word Embedding and ChatGPT

Don't be misled by the heading of this section. There does not seem to have a way for Excel to successfully simulate word embedding (also known as text embedding), let alone to simulate ChatGPT. However, since ChatGPT has brought us a dizzying AI revolution which may reshape the course of human as an intelligent species, I would like to talk about them briefly here.

As mentioned before, to process text data via computers, we need to arrange the data in a table format. This is exactly what word embedding does. The brilliance of embedding is to use a numerical vector to define a word, each numerical value representing a certain unknown property of the word. Thus, all words in a corpus can be represented numerically by a matrix. For example, assume our corpus is the sentence "At the most beautiful place remember the most beautiful you" and every word has seven properties; once we have tokenized each word, we may randomly get a matrix like Table 14-1.

Table 14-1. *A sample word embedding matrix. Notice that some words are*
repeated

Words	P1	P2	P3	P4	P5	P6	P7
at	0.6	-0.9	0.1	-0.4	-0.2	-0.1	-0.9
the	0.2	-0.9	1.0	0.9	-0.8	0.5	0.0
most	0.5	0.5	0.4	-0.1	-0.1	0.2	0.9
beautiful	0.0	0.9	0.3	-1.0	0.7	-0.4	-0.3
place	-0.6	0.7	-1.0	-0.9	0.0	-0.3	1.0
remember	0.8	1.0	0.8	0.5	0.0	-0.5	-0.4
the	0.2	-0.9	1.0	0.9	-0.8	0.5	0.0
most	0.5	0.5	0.4	-0.1	-0.1	0.2	0.9
beautiful	0.0	0.9	0.3	-1.0	0.7	-0.4	-0.3
you	0.6	-0.1	0.5	-0.4	0.0	0.8	-0.5

By convention, when a word is represented by a vector of N numerical values, we say that the embedding dimension size is N, or the dimension of the word vector is N, instead of saying N properties. In natural language processing (NLP), especially for a large language model (LLM), the embedding dimension size is usually at the magnitude of hundreds, thousands, or even much larger.

When we conducted sentiment analysis on hotel reviews, we predefined stop words and polarity words. However, in word embedding, the numerical vector of each word is not predefined; instead, it is trained and optimized through a neural network. This is the beauty of the embedding technique! In a LLM, the first hidden layer is always the embedding layer, which reveals that embedding is the foundation of very large language models. Once the training and testing of such a model is completed, the embedding matrix is optimized, that is, pretrained. A successfully well pretrained embedding matrix can precisely define almost all the words for other language processing tasks, for example, the famous Word2Vec.

Assume that our neural network model achieved a well pretrained embedding matrix whose word vector is of two dimensions. Our embedding matrix might give us (king: -0.2, 0.9), (man: -0.2, 0.5), (woman: 0.3, -0.2), and (queen: 0.3, 0.2). We then could mathematically "prove" that king – man + woman = queen, an interesting equation originally obtained by Word2Vec (note, Word2Vec's embedding dimension is unknown, but roughly between 100 and 1000).

GPT stands for Generative Pre-Trained Transformer. You may get confused by the two words "generative" and "transformer". Well, the data mining methods or machine learning models we have learned in this book can only make discoveries, they do not generate anything new by themselves. A GPT model, however, can make creations, that is, "generate", by themselves. Transformer is a new type of neural network model that can learn and evolve. In some sense, it has self-attention. At the end, ChatGPT is a large language model developed by the company OpenAI and pretrained to perform various NLP tasks.

The embedding technology plays a very important role in ChatGPT. Though OpenAI released that ChatGPT 3 has 175 billion parameters, its embedding dimension is unknown (at least to me). But we can safely say that its embedding dimension is fairly large.

Reinforcement Exercises

The only reinforcement exercise is presented in the file Chapter14-HW.xlsx. Inside this file, you shall find out that a dictionary has already been prepared for you, though you are welcome to revise it. Based on the given dictionary, Chapter14-HW-withAnswers. xlsx presents a standard solution without making use of the TEXTJOIN function. As its name suggests, Chapter14-HW-withAnswers-UseTEXTJOIN.xlsx provides a solution that makes use of the TEXTJOIN function. At the time when I revise this book, I am sure that TEXTJOIN function is available to most Excel users.

Review Points

1. Structured and unstructured data

2. Applications of text mining

3. Term frequency, polarity score, tokenize, keywords, stop words, stemming, n-gram

4. Excel functions IF, COUNTIFS, INDEX, MATCH, AND

5. Excel functions IFERROR, OR, ISBLANK, MID, TRIM, LEN, LOWER

6. Name Manager

7. Text to Columns

8. Excel function TEXTJOIN

9. Word embedding

CHAPTER 15

After Excel

Excel is a remarkable tool for business intelligence, office jobs, and many other applications. Its greatness comes from many features it has, including formula and functions, autofill, automatic workbook calculation, add-in, etc. Via automatic workbook calculation, any modifications on the formula or the value inside a cell can trigger recalculations of all other cells that are directly or indirectly related to this cell, even if they are on a different worksheet. The recalculation can also reshape charts that are dependent on these cells. The value of a cell can also be linked to a form control, which renders visual control of data values and charts.

Autofill automates the calculation of a large number of cells, which is an extremely powerful feature. This feature makes Excel capable of working with a dataset of large size. Data in Excel is automatically arranged in a table format, making Excel naturally suitable for many data mining techniques, too.

The hundreds of functions in Excel make Excel programmable. We can treat an Excel formula like a programming statement, a worksheet like a modular function, a class, or a program. The available functions can be categorized based on their specific purposes, for example, financial, mathematical, statistical, references and lookup, etc. The creative uses of these functions offer us the opportunities to learn data mining step by step through Excel.

The power of Excel has been evolving all the time. In the first Excel version I used, Excel 2003, its available numbers of the rows and columns were 65,536 and 256, respectively. Since version 2007, the number of available rows is more than 1 million and the number of available columns is more than 16,000. The capacity of today's Excel allows it to operate on large datasets.

The available number of functions of Excel has been increasing all the time, too. Table 15-1 lists the new functions since version 2010 and the total functions since version 2007. Data in Table 15-1 were compiled from https://support.office.com/. Generally speaking, a larger set of functions indicates greater capabilities.

© Hong Zhou 2023
H. Zhou, *Learn Data Mining Through Excel*, https://doi.org/10.1007/978-1-4842-9771-1_15

I am sure by now we have accumulated quite some understanding about data mining and won't consider data mining or machine learning a mysterious technology. We may also be asking what else Excel can do as there are many other data mining methods we have not touched in this book, for example, random forest, support vector machine, principal component analysis, deep learning, natural language processing (NLP), etc.

Table 15-1. *Number of functions in different Excel versions*

Office Version	2007	2010	2013	2016	2019	2023//365
New Functions		56	51	6	16	11
Total Functions	350	406	457	463	479	490

I would like to emphasize that the purpose of this book is to introduce basic data mining methods through Excel examples. I have no intention to cover all possible data mining methods in this book, though the simplified versions of some other data mining methods can also be illustrated through Excel examples.

We must also admit that Excel is not a programming language, that is, we must admit the limit of Excel in learning data mining. In the last chapter, I mentioned that Excel is not suitable for certain text mining tasks. In Chapter 12, I could show how to conduct an association analysis with itemsets of size 2 only. When the itemset size increases, the number of different itemsets goes up exponentially, and it becomes unsuitable and eventually impossible to conduct association analysis through Excel.

Originally, data mining software tools provided semiautomatic mining processes. Users must apply one data mining method at a time. To compare the performances of different methods, users must try out different methods and compare their performances manually. Today automatic data mining becomes more and more popular. Software tools such as RapidMiner provide "Auto Model" that allows users to apply as many mining methods as they select and can automatically compare the performances of these methods. Some Python packages such as AutoGluon can even automatically complete the EDA and feature selection processes, generate new features (e.g., applying the embedding technique on text data), and construct an ensemble machine learning model that consists of a set of selected data mining methods. Certainly, it is much more difficult for Excel to provide such an automatic data mining feature, even though the

Azure machine learning add-in is quite powerful. Once we have learned the data mining knowledge through Excel, we do need to be skilled in data mining software tools and/or languages if we have the desire to become a data mining or machine learning expert.

Automatic data mining requires minimum effort in the mining phase but pushes large effort into the data preparation phase. As mentioned before, Excel can provide users direct visual examination of every data preparation operation. Thus, Excel will always play an important role for data mining practices. When the dataset is too large for Excel to operate on, Excel can still be a good tool to test the data preparation process at a smaller scale.

Learning is a step-by-step process; it is not and never will be an automatic process. We know that learning by doing is the best educational approach. Learning data mining using Excel forces learners to go through every data mining step manually, offering them an active learning experience. I would state that the automatic data mining trend makes Excel even more important in studying data mining or machine learning.

Index

A

Activation function, 226, 230
Agglomerative clustering, 122
Artificial intelligence (AI), 5, 272
Artificial neural network, 225, 250
    (*see also* Neural network)
Association analysis
    concept of support, 211, 212
    confidence, 212
    excel, 222–224
    instructions, 212
    itemsets of size, 213, 214
    market basket analysis, 211
    reinforcement exercise, 222
    shopping basket data, 213
    support percentage, 211
AutoGluon, 278
Automatic data mining, 278, 279

B

"Big Data", 200

C

Case transformation, 253, 254
Categorical attributes, 191, 193, 197,
    205, 206
ChatGPT, 6, 272–274
Clustering
    agglomerative clustering, 122
    data point, 37

divisive clustering, 122
    record, 37
    subject, 37
    types, 37
Credit card
    company, 104
Cross-validation
    excel
        autofill, 74
        cells, 74
        Chapter5-cv-1c.xlsxf file, 72
        instructions, 73
        inter-group variance, 75
        iris dataset, 72
        LDA, 72, 73
        LINEST function, 72
        parameter set, 73–76
        random algorithm, 72
        Solver, 75
        testing dataset, 75, 77
        worksheets, 72–74
    holdout method, 71
    K-fold, 72
    K values, 108
    LOO, 72
    reinforcement exercises, 84
    rest, 77
    training dataset, 77

D

Data cleaning, 187, 196–200
Data engineering, 5

281

© Hong Zhou 2023
H. Zhou, *Learn Data Mining Through Excel*, https://doi.org/10.1007/978-1-4842-9771-1

I, J

K

Printed in the United States
by Baker & Taylor Publisher Services